# The Moon

## Myths, Mystics, and Missions

First Edition: 2026

Author (Pen Name): Balasaheb Pendharkar

Legal Name: Bhushan Kailas Pendharkar



For permissions, bulk orders, or feedback:

**Email: vruttachitradocumentaries.book@gmail.com**

## Disclaimer

This book is intended for educational and informational purposes only. While every effort has been made to ensure accuracy, the content may include interpretations, historical narratives, cultural beliefs, and speculative discussions.The author does not claim absolute scientific or historical authority. Readers are encouraged to consult primary academic sources where required. Any resemblance to real persons or events is coincidental.

Some images and illustrations in this book are generated or assisted using artificial intelligence tools. These visuals are original creations used for representational purposes and do not depict real persons, events, or proprietary material.

*To every mind that looked at the Moon*

*not as an object—*

*but as a question.*

# TABLE OF CONTENTS

## PART I: Birth of the Moon – The Celestial Beginning



## PART II: Moon in Prehistoric and Primitive Cultures



## PART III: Ancient Civilizations and the Moon



## PART IV: Moon and the Calendar – Measuring Time



## PART V: Moon in Astrology, Spirituality, and Mysticism

## PART VI: Moon in Folklore, Literature, and Art



## PART VII: Moon in Medieval Science and Observation



## PART VIII: The Moon and Modern Space Exploration



## PART IX: Moon and Earth – Environmental & Biological Connections



## PART X: Future of the Moon – Colonies, Mining & Mysteries

PART – I

# Birth of the Moon

## THE CELESTIAL BEGINNING

# 1-The Giant Impact Hypothesis

## A Violent Beginning

***"In the infancy of our solar system, when chaos reigned and fire ruled, a violent collision would change Earth's destiny forever—and gift it with a Moon."***

### A Solar System Born in Chaos

Let's rewind time—**4.5 billion years ago.**

The Sun had just ignited, lighting up the young solar system like a cosmic bonfire. Planets weren't planets yet—just hot, molten balls of rock hurtling through space, crashing, colliding, and reforming.

Earth was still in its infancy. And somewhere, spinning silently in its own orbit, was a rogue planet—**Theia.**

She was roughly the size of Mars. Not a villain, not a hero. Just another wandering body in a violent solar system. But fate—if we dare call it that—was about to make her the most important character in Earth's story.

### The Great Collision

One day (or what passed for a day back then), Theia and the proto-Earth met—not gently, but with apocalyptic force.

A side-glancing impact hurled molten rock and debris into orbit.

Temperatures soared, metals vaporized, crust shattered.

Earth survived… barely. Theia did not.

But from that wreckage — a miracle.

The fragments of Earth and Theia, locked in a furious orbital dance, began to clump together. Over time, gravity sculpted them into a single sphere: **the Moon.**

It wasn't a gentle birth. It was forged like a sword—**in chaos and flame**.

What Makes This Hypothesis So Believable?

It sounds cinematic. But this isn't fiction. This is the Giant Impact Hypothesis—the leading scientific theory of how the Moon came to be. And it's backed by some compelling facts:

#### 1. Earth and Moon: Isotopic Twins

Samples from Moon missions show that oxygen and titanium isotopes in lunar rocks are nearly identical to Earth's. That's bizarre… unless both were once part of the same body.

#### 2. Missing Iron

The Moon is strangely low on iron. If it were a captured asteroid or sibling planet, it should've had a core. But if it was born from Earth's outer layers? That makes sense.

#### 3. A Tilted Tale

The Earth-Moon system's tilt, spin, and momentum are best explained by a massive, off-center collision. It accounts for Earth's 23.5° tilt—crucial for seasons, climate, and life.

### Theia: The Forgotten Planet

Theia didn't vanish completely. Some scientists believe her remains might still be buried inside Earth's mantle—giant iron-rich blobs detected deep beneath Africa and the Pacific.

Imagine that. The Moon in the sky… and Theia's ghost still inside us.

### But Not Everyone Agrees

As elegant as it is, the Giant Impact theory has some gaps:

If Theia was a separate planet, shouldn't the Moon's isotopes be slightly different?

If the impact was that hot, why does the Moon still contain volatile elements like water?

To fill these holes, scientists proposed tweaks:

Maybe Theia and Earth formed in the same zone, making them chemically alike.

Or maybe Earth turned into a "synestia"—a donut-shaped cloud of magma and vapor—from which the Moon condensed.

Some even suggest multiple smaller impacts, not one big one.

Science, after all, is a living story—not a fixed script.

### Echoes of Mythology

Here's what's beautiful: ancient myths echo this chaos-born origin story in strange and poetic ways.

In Greek mythology, Theia was the mother of Selene, the Moon goddess.

In Hindu cosmology, the Moon (Chandra) is churned from the cosmic ocean, a product of conflict and magic.

In Babylonian lore, the heavens are formed when the god Marduk slays the dragon Tiamat—a chaotic celestial battle, not unlike Theia's collision.

They didn't have telescopes. But they sensed something: the Moon wasn't just created… it was forged.

### Why It Matters

This isn't just ancient history.

The Moon controls tides, stabilizes our axis, and may have even been key to the evolution of complex life on Earth.

Without the Moon, Earth's climate might have wildly swung, destabilizing early ecosystems.

Tidal pools, stirred by lunar gravity, may have nurtured the first life.

And today, the Moon is our first step toward becoming a multi-planetary species.

Understanding its origin isn't just about the past. It's about where we go next.

### Final Thoughts: A Broken World, a Beautiful Result

"The Moon is a child of destruction. A scarred memory of a violent past. Yet it became Earth's greatest companion, its gentle nightlight, and its cosmic stabilizer."

Every time you look up at the Moon, you're staring at a graveyard… and a cradle.

A piece of Earth that was lost—then reborn.

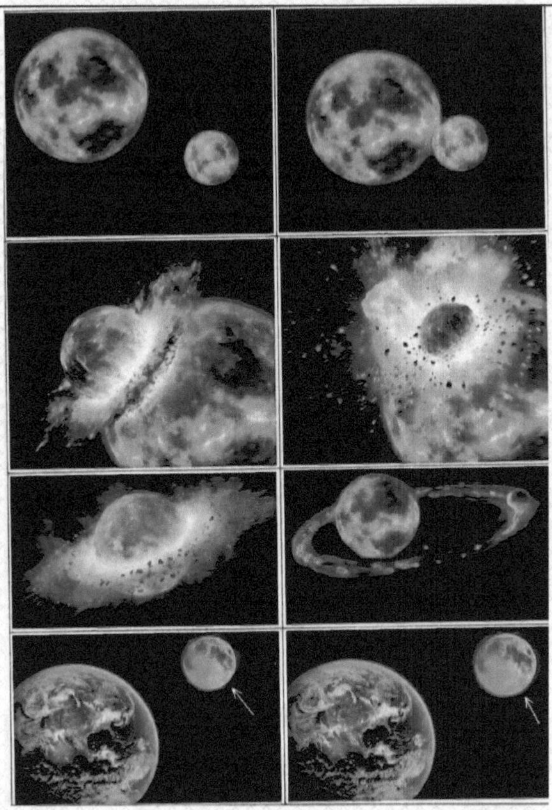

1. 4.5 billion years ago The Proto - Earth and The Theia just hot, molten balls,

2. a rogue planet—Theia , Moving in Chaotic Orbit.

3.The Theia with Earth in angular collision.

4 impact hurled molten rock and debris into orbit

5 & 6. The fragments of Earth and Theia, locked in a furious orbital dance, began to clump together

7. Early Moon And Earth

8. Finally the Moon as we seen in present day.

# 2 -Prehistoric Perceptions Of The Moon

*"Before science, telescopes, or written words, humans looked up and saw a glowing mystery — and began to dream."*

## The First Mystery in the Sky

Long before history was written, before villages became cities, before kings or priests, there were only fire-lit nights and star-filled skies. Among all the lights that burned above, one stood apart — **the Moon.** Unlike the distant stars, the Moon changed. It waxed, it waned, and it disappeared and reappeared. It seemed alive.

To our prehistoric ancestors, the Moon was not just a light in the sky — it was a companion, a clock, a calendar, and above all, a source of awe.

## What Did Our Ancestors See?

When you stand under a sky without city lights, the Moon feels overpowering. In prehistoric times, it shaped how humans hunted, gathered, celebrated, and even told stories around campfires.

- The Changing Face: Every few nights the Moon looked different — crescents, halves, full, then gone. This cycle became the earliest rhythm humans noticed apart from the Sun's day-night pattern.
- Predictable Yet Mysterious: Unlike storms or animal migrations, the Moon's phases repeated with precision. This predictability allowed early humans to count time in a world without clocks.

## Moon in Stone Age Life

### 1. Lunar Calendars Before Time Itself

Archaeological discoveries show that Stone Age people tracked the Moon long before they tracked the Sun.

- Lascaux Caves (France, 17,000 years old): Painted alongside animals are dots and lines that some researchers believe may represent lunar cycles.

- The Blanchard Bone (France, 30,000 years old): A piece of mammoth ivory carved with 69 notches — interpreted by scholars as recording the days of a lunar cycle.

These are perhaps the earliest calendars humanity ever made — not to predict harvests yet, but to track cycles of hunting, animal migrations, and seasons.

#### 2. Moon and Menstruation

One of the most striking connections our ancestors may have noticed was between the lunar cycle (~29.5 days) and the menstrual cycle (~28–30 days). This parallel gave the Moon a deeply feminine, sacred status. The Moon became associated with fertility, womanhood, and the mystery of creation.

### Symbolism and Spirituality

#### 3. The Moon as Rebirth and Immortality

The Moon's cycle of waxing, waning, and returning from darkness mirrored the human experience of life, death, and rebirth. For prehistoric people, this was evidence that nothing truly dies, it transforms. The Moon disappearing at Amavasya and returning again became a cosmic guarantee of renewal.

#### 4. Moonlight and the Spirit World

The full Moon transformed the night. Animals behaved differently, humans stayed awake longer, and rituals were performed. Across cultures, full moons were thought to open portals to the spirit world. Shamans used these nights for trance-dances, healing, and contacting ancestors.

Moonlight, with its cold silver glow, was seen not as sunlight's reflection, but as a mystical energy in its own right — capable of inspiring visions and guiding souls.

### Megaliths, Monuments, and Moon-Watching

#### 5. Stonehenge and the Moon

While Stonehenge (England, ~3000 BCE) is often linked with the Sun, some of its stones align with lunar standstills (the extreme rising and setting points of the Moon, occurring every 18.6 years). This suggests ancient sky-watchers were keenly observing long-term lunar cycles.

**6. Nabta Playa, Egypt (~6000 BCE)**

Before the pyramids, desert tribes erected stone circles that may have been used as ceremonial calendars aligned with both solar and lunar events. These early astronomer-priests were mapping the Moon's role in marking ritual times.

Such monuments tell us that the Moon was not just noticed — it was measured, celebrated, and integrated into sacred landscapes.

## Moon Across Early Cultures

The universality of lunar fascination shows how deeply it shaped the human imagination.

Africa: The Moon was tied to water and fertility — both essential for survival. In some myths, the Moon controlled rainfall and rivers.

Australia (Aboriginal Dreamtime): The Moon was a man who died and was reborn each month, explaining its phases. In some regions, he was a lover chasing the Sun.

South America: Ancient rock carvings (petroglyphs) show crescents and full circles, often linked with planting and fertility rituals.

India: Rock carvings and ritual objects from early Indus Valley cultures suggest lunar worship long before the Vedas were written.

Wherever humans lived, they wove the Moon into myth, ritual, and survival practices.

## The Moon as the First Clock

Before sundials, before hourglasses, before mechanical clocks, there was the Moon.

- Hunters could plan drives and expeditions under bright full moons.
- Gatherers knew which fruits ripened according to seasonal moon cycles.
- Early farmers eventually tied sowing and harvesting to lunar phases.

The Moon was humanity's first shared calendar. It provided rhythm to an otherwise chaotic and uncertain world.

### Storytelling, Art, and Ritual

The Moon was more than a clock — it was the first muse.

- Cave paintings may have encoded its cycles.
- Ritual dances imitated its crescents and rounds.
- Myths were told of lovers, gods, and beasts tied to its phases.

Around communal fires, the Moon became a canvas for imagination: a face in the sky, a rabbit, a woman, or a hunter. Each culture filled its silver glow with meaning.

Primitive Humans Watching the Full Moon

## Final Thoughts: The First Muse

"When there was nothing but stars and firelight, the Moon was our mirror, our map, our myth."

For prehistoric people, the Moon was not a distant object but an intimate companion. It regulated time, inspired art, symbolized birth and rebirth, and offered a link between earth and spirit.

The fascination that began in caves and stone circles still lingers today, in poetry, science, and dreams of space travel. In many ways, our journey to the Moon in 1969 was not the beginning of exploration — it was the culmination of humanity's oldest love story with the night sky.

Hunter-Gatherers Planning by Moonlight

# 3- Ancient Geological Evolution Of The Moon

## From Fire To Crust

*"Before it became a silver disc in the night sky, the Moon was a world ablaze—a molten orb shaped by fire, impacts, and cosmic forces."*

### The Birth of a Magma Ocean

After the giant impact with Theia, the newborn Moon wasn't solid. It was an ocean of molten rock—a magma ocean, glowing red against the darkness of space.

Temperatures soared past 1600°C, enough to melt every mineral.

For millions of years, the Moon churned like a cosmic cauldron. Heavier materials like iron and nickel sank inward. Lighter ones—like anorthosite and plagioclase—rose to form a crust.

This separation of elements, known as differentiation, gave the Moon its earliest geological identity.

### The Lunar Highlands: Ghosts of the First Crust

As the magma ocean cooled, it began to crystallize.

Floating minerals, mostly light-colored plagioclase feldspar, formed a rigid crust—what we now call the lunar highlands. They are still visible today: the bright, rugged areas on the Moon's surface.

These highlands are older than any Earth rock, dating back over 4.3 billion years—a preserved record of the solar system's ancient furnace.

### The Great Bombardment: Fire from the Heavens

Between 4.1 to 3.8 billion years ago, the Moon faced its greatest test—the Late Heavy Bombardment.

Thousands of asteroids and comets, leftovers from planetary formation, smashed into its surface.

Massive craters like Imbrium, Serenitatis, and Orientale were born. Lava flooded the impact basins. A planetary punching bag, the Moon wore every scar.
The Earth was hit too, but erosion and tectonics erased most of that history. **The Moon remembers.**

### Lunar Maria: The Moon's Dark Seas

In the aftermath of the bombardment, the Moon's insides still simmered.

Molten basalt erupted through cracks and flooded the lowlands.

These lava plains, now known as **maria** (Latin for "seas"), cooled into dark, flat regions visible from Earth. They include Mare Tranquillitatis, Mare Imbrium, and Mare Serenitatis.

They're not water—but ancient seas of fire, frozen in time.

### Volcanism and Tectonics: A Brief Flame

For hundreds of millions of years, volcanic activity continued—though never as violently as Earth's.

Some ancient domes and lava tubes still dot the Moon's surface. But tectonics? Almost absent.

The Moon lacked the internal energy to form plate boundaries. Instead, its surface evolved slowly, through cooling and bombardment.

By about 1 billion years ago, the Moon had gone geologically silent.

### The Quiet Surface: Regolith and Micrometeorites

With no atmosphere to shield it, the Moon endured billions of micrometeorite impacts, breaking down rocks into a dusty layer called **regolith**.

This fine, gray powder blankets the surface and holds the stories of eons. It's why astronaut footprints remain intact—they're embedded in history.

### Lessons for Earth

Studying the Moon's geology is like flipping through the oldest pages of Earth's lost diary.

Since the Moon never had weather or plate tectonics, its surface is like a **cosmic fossil record**—untouched, raw, and incredibly revealing.

It tells us how planets form, how magma oceans evolve, and how worlds die—or endure.

### Final Thoughts: A World Etched in Stone

"The Moon is not just a mirror to Earth's sky—but to its soul. Where Earth heals, the Moon remembers. Where Earth grows, the Moon preserves."

Every ridge, crater, and sea on the Moon is a monument to time itself.

**It is not dead. It is eternal.**

# 4-The Moon's Role In Earth's Tides And Stability

*"Without the Moon, Earth would still spin. But life… life as we know it, might never have begun."*

### The Invisible Connection

Stand at the edge of the ocean and watch the waves roll in. Twice a day, the waters rise and fall — like the breath of the planet. But what breathes life into this rhythm?
**The Moon.**

Though it hangs 384,400 kilometers away, the Moon's gravitational pull silently sculpts Earth's most dynamic systems — from ocean tides to planetary stability.

It is both guardian and choreographer of Earth's balance.

### The Physics Behind the Pull

The Moon pulls on Earth — and Earth pulls back. But water, being fluid, responds more dramatically.

- On the side facing the Moon, the ocean bulges outward due to direct gravitational attraction.
- On the opposite side, another bulge forms — caused by inertia, as Earth is pulled slightly away from the water.

These bulges rotate with Earth's spin, creating two high tides and two low tides each day.

This tidal motion is more than a beach-time curiosity — it is a planetary heartbeat.

### Tidal Friction: The Moon is Slowing Earth Down

Here's a twist: the Moon is slowing down Earth's rotation.

As tides move across the globe, they create friction against the ocean floor. This tidal friction acts like a brake on Earth's spin, slowly increasing the length of a day.

400 million years ago, a day was only about 22 hours.

In another 100 million years, it could be 25 hours.

And this friction isn't free — the Moon pays the price.

Every year, it moves 3.8 cm farther away from Earth. Our cosmic companion is drifting… slowly but steadily.

**Climate and Axial Stability**

One of the most critical functions the Moon performs isn't visible — it stabilizes Earth's axial tilt.

Without the Moon, Earth's axis could wobble wildly — from 0° to 85° — like a spinning top without a base.

This would cause catastrophic climate shifts — poles might become tropical, while equators could freeze. Life would struggle to adapt to such chaos.

With the Moon in place, Earth's tilt stays near 23.5°, allowing predictable seasons, climates, and ecosystems.

**The Spark of Life?**

Tidal zones — the shifting edges where land meets sea — may have been the cradle of life.

In prehistoric times, these wet-dry cycles, powered by lunar tides, may have concentrated organic molecules, allowing them to link and evolve.

RNA strands, the first self-replicating molecules, may have formed in these environments — birthing the first life on Earth.

So in a way, we may all be descendants of a tide pool stirred by the Moon.

**The Moon's Effect on Wind and Weather**

Lunar gravity not only tugs at oceans, but also at Earth's atmosphere.

Though less intense, the Moon causes atmospheric tides — slight shifts in air pressure that subtly influence wind patterns, jet streams, and possibly monsoons.

In delicate climate systems like India's, even a minor lunar push can nudge seasonal behavior.

### Impact on Human Systems

- Calendars: Lunar cycles influenced the earliest calendars — from Babylon to India to Mesoamerica.
- Agriculture: Farmers observed moon phases for sowing and harvesting — a practice still alive in rural traditions.
- Navigation & Rituals: Ocean tides guided sailors. Full moons were linked to feasts, rituals, even crime spikes.

The Moon became a timekeeper, a myth-maker, and a meteorologist.

### What Happens If the Moon Disappears?

Science fiction loves this scenario. But the reality?

- Tides would shrink to 30–40% of current size (only influenced by the Sun).
- Earth's axial wobble could destabilize — turning Earth into a climatic roulette.
- Days would no longer lengthen — Earth's spin would remain chaotic.
- Life, if it survived, would evolve in very different ways.

The Moon's absence would create a world alien to us.

### Final Thoughts: A Hidden Architect

"The Moon does not just decorate the night sky — it designs Earth's rhythms."

From deep-sea currents to the whisper of wind in wheat fields, from the ebb of waves to the cycle of seasons — the Moon is everywhere, even when invisible.

To look at the Moon is to see the architect of balance — the unseen force that makes Earth habitable, fertile, and steady.

We owe more than poetry to that pale disc in the sky.

We owe our very lives.

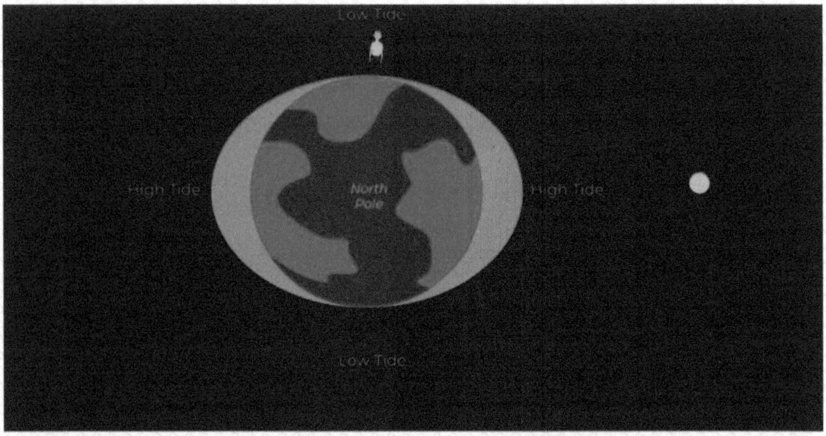


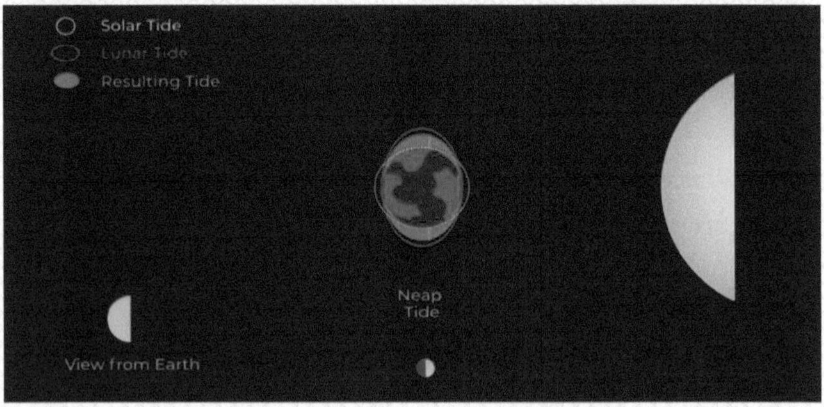

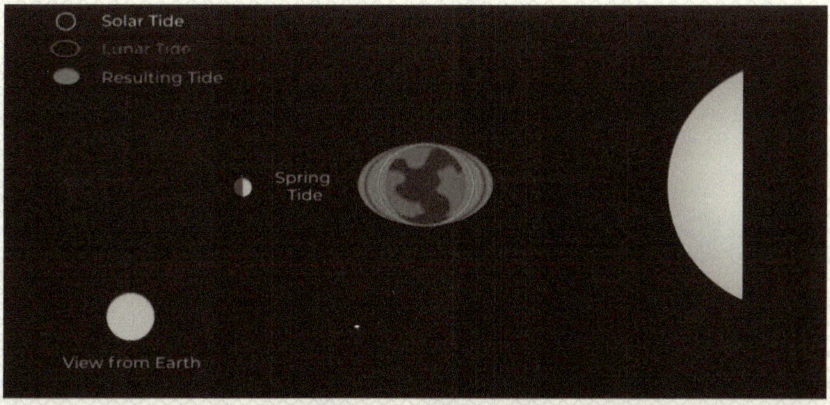

**Courtesy : NASA**

# 5- Lunar Influence On Biological Rhythms - (Circadian & Reproduction)

*"We've always known the Moon affects the tides… but what if it also tugs at the tides within us?"*

**Introduction: The Moon's Subtle Power Over Life**

From ancient healers to modern chronobiologists, the idea that the Moon affects living organisms has intrigued humanity for millennia.
It isn't just folklore. Science now reveals that the Moon's gravitational pull, light cycle, and periodicity subtly influence biological clocks, reproductive timing, and even behavior — not only in humans but across the animal kingdom.

**Circadian vs. Circalunar Rhythms: The Body's Timekeepers**

All life on Earth is tuned to natural cycles:

- Circadian rhythms follow a 24-hour solar day. They regulate sleep, metabolism, body temperature, and hormone secretion.
- Circalunar rhythms, less talked about, follow the 29.5-day lunar cycle. These rhythms are especially strong in marine life, insects, and some mammals, including humans.

Our internal clocks — governed by the brain's suprachiasmatic nucleus — are fine-tuned by external cues: daylight, temperature… and yes, moonlight.

**Marine Life: Born by the Moon**

Many marine species synchronize their mating and spawning with the lunar phases.

- Palolo worms in the Pacific rise en masse to the ocean surface during the last quarter of the Moon to reproduce.
- Grunion fish ride high tides after full moons to lay eggs on beaches.

- Coral reefs release reproductive cells in mass spawning events during full moons.

In these ecosystems, the Moon is not just a clock — it's a fertility goddess.

### Plants and Lunar Cycles

Some research suggests plant growth, sap flow, and germination rates vary with moon phases.
Certain species show enhanced leaf expansion during waxing phases, possibly responding to gravitational effects or nocturnal moonlight.

Farmers have long used lunar planting calendars, observing better yield when seeds are sown during waxing moons — though the science is still debated.

### Animals and Moonlight

Wildlife across the globe responds to the lunar cycle:

- Predators hunt more actively during full moons due to increased visibility.
- Prey animals reduce activity under bright moonlight to avoid detection.
- Some bats avoid hunting during full moons — a phenomenon called lunar phobia.

Even night-singing birds like nightjars and whip-poor-wills increase vocal activity during moonlit nights — possibly for mating.

### Human Biology and the Moon

This is where myth meets mystery.

Menstrual Cycles

The average menstrual cycle (~28 days) closely mirrors the lunar month (~29.5 days). In historical and tribal communities, stronger synchronization has been reported — possibly due to natural light exposure.

Some researchers speculate that **moonlight**, particularly during the **full moon**, may have once helped synchronize menstrual and ovulation cycles across groups.

This may have aided **communal child-rearing** and enhanced **tribal cohesion**.

### Sleep and Moon Phases

Modern studies using sleep labs have found:

- Lower melatonin levels, reduced deep sleep, and delayed sleep onset during full moons.
- Even in dark rooms with no visible moonlight, participants' sleep changed — suggesting a possible endogenous lunar sensitivity.

The mechanism isn't fully understood — but it's real.

### Moon and Mental Health

Historically, full moons were blamed for madness, giving rise to the term "lunatic."

Some emergency room studies show slight increases in seizures, aggression, mood swings, and psychiatric admissions during full moonsHowever, the evidence is **inconclusive** — suggesting cultural factors may exaggerate natural fluctuations.

### Reproductive Timing and Lunar Influence

Some studies show slight statistical bumps in:

- Conceptions around full moons
- More natural births around new moons or full moons
- Fertility hormone cycles potentially linked to moonlight exposure

In animals like horses, rabbits, and some primates, mating behaviors increase around full moons.

### The Moon as a Timekeeper of Life

Whether it's spawning coral, sleeping humans, or flowering plants — the Moon continues to act as a cosmic metronome.

### Final Thoughts: The Moon Within Us

"The Moon doesn't just orbit above us — it whispers through our blood, our sleep, our generations." We are lunar creatures, sculpted not only by sunlight but by silvery nights.

In every heartbeat that follows a rhythm, in every cycle of rest and renewal, the Moon's quiet influence lingers — unseen, but deeply felt.

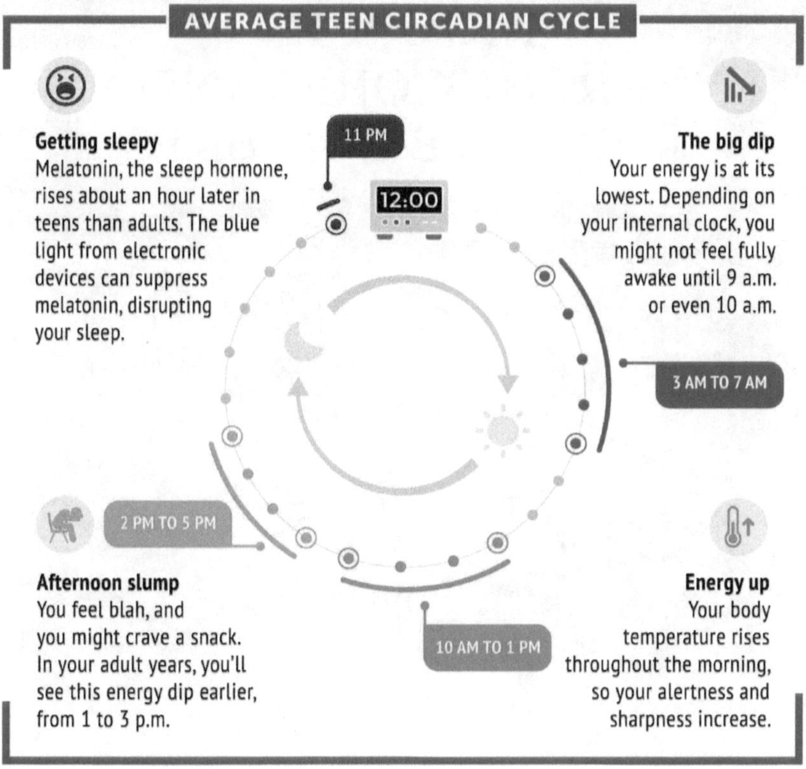

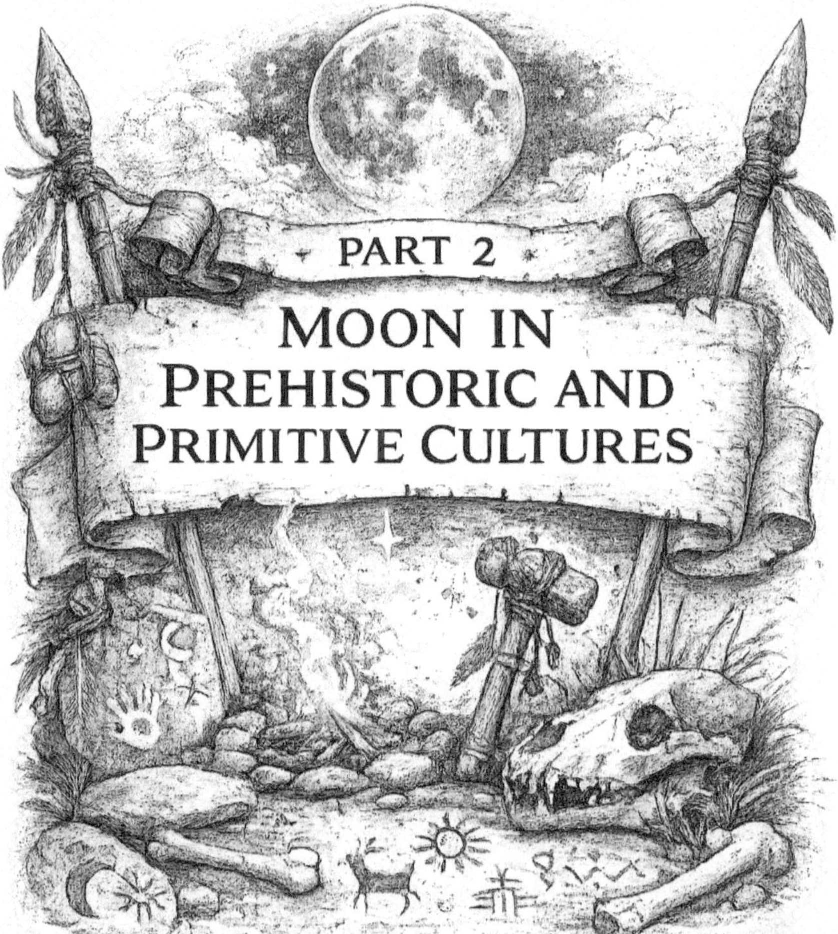

# PART 2
# MOON IN PREHISTORIC AND PRIMITIVE CULTURES

# 6- Moon In Cave Art And Stone Circles

***"Before alphabets, before temples, before myths — humans painted, carved, and aligned the Moon into stone and memory."***

**The First Lunar Artists – Prehistoric Cave Art**

Long before humans built cities or recorded history, they observed the sky.

Caves across Europe, Africa, and Asia contain primitive paintings that date back tens of thousands of years. While most feature animals and hunting scenes, certain marks and patterns hint at lunar observation.

**Notable Findings:**

- La Pasiega Cave (Spain) and Chauvet Cave (France) show dot patterns and cycles that some researchers believe represent lunar phases.
- In Lascaux Cave, a mysterious cluster of dots near a bull painting has been interpreted as a lunar calendar — possibly a record of Moon-based timekeeping linked to hunting seasons.

**Interpreting the Dots:**

- Repeating 13 or 29 dots are seen as moon cycles (e.g., the 29.5-day lunar month).
- Some arrangements could represent the waxing and waning phases.
- These symbols suggest that prehistoric humans were tracking time by the Moon — long before written calendars.

**Why the Moon?**

The Moon would have been the brightest night object for our ancestors.

Its predictable patterns likely helped early humans anticipate seasonal changes, track animal migrations, and organize fertility or hunting rituals.

Lunar visibility in the sky made it a natural sky-clock for nomadic life.

**Stone Circles and Lunar Alignments**

Beyond cave walls, humans started to sculpt the Moon into landscapes.

Across Europe, Asia, and even parts of Africa, megalithic structures — giant stone circles, mounds, and alignments — were built to mark celestial events.

Not Just for the Sun:

- Callanish Stones (Scotland): Aligned to the 18.6-year lunar cycle (known as lunar standstill).
- Newgrange (Ireland): While primarily solar, may also have had lunar ceremonial roles.
- Nabta Playa (Egypt): A prehistoric stone circle that aligns with both lunar and solar cycles — possibly one of the oldest observatories (7,000+ years old).

### The Lunar Standstill Mystery

Every 18.6 years, the Moon reaches its extreme rising and setting points — this is called a major lunar standstill.

At places like Callanish, certain stones align perfectly with the Moon's lowest and highest points in the sky during this cycle.

This means ancient builders understood long-term lunar motion — a staggering achievement.

### Ritual, Myth, and the Night Sky

These structures weren't just calendars — they were temples without roofs.

People likely gathered during solstices, equinoxes, and full moons for rituals, dances, burials, or fertility ceremonies.

Many stone circles are found near burial sites, suggesting a belief in life, death, and rebirth — all tied to lunar cycles.

### A Shared Legacy Across Cultures

While the exact meanings of these sites are lost, the global pattern is clear:

- The Moon mattered.
- Time was sacred.
- Sky and stone were connected.

From the painted caves of France to the windswept circles of Scotland and the sun-scorched plains of Africa, the Moon was the first calendar, goddess, and mystery.

### Final Thoughts: Moon as Muse and Marker

"Before gods had names, before time had clocks, the Moon marked our beginnings."

The Moon wasn't just seen; it was experienced, counted, and carved into the very bones of the Earth.

Even now, when we stand among ancient stones or gaze at prehistoric symbols, we're standing at the crossroads of cosmic awe and human curiosity.

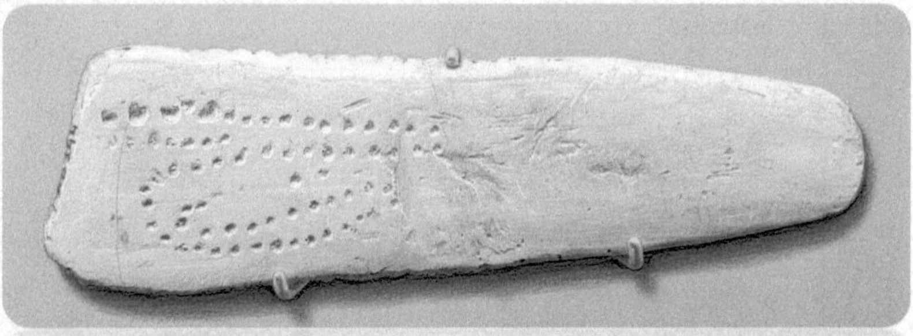

**Image courtesy:** [Don Hitchcock], *Wikimedia Commons.*

# 7 - Lunar Cycles And Fertility Worship

*"Long before science explained ovulation or gestation, the Moon pulsed through wombs and rituals, guiding generations with its glow."*

**The Mysterious Connection Between Moon and Woman**

Across ancient cultures, one belief echoed again and again: the Moon governed fertility.

And it wasn't without reason:

•The average menstrual cycle is ~28-29.5 days—remarkably close to the lunar month.
• Both cycles influence moods, energies, fluids, and biological changes.
• Ancient people noticed this harmony—and revered it.

For them, the Moon wasn't a rock. It was a womb in the sky.

**Sacred Femininity and the Moon**

In matriarchal societies and goddess-worshiping cultures, the Moon was feminine — ever-changing, nurturing, powerful.

She was:

• The maiden in the waxing crescent.

• The mother in the full moon.

• The crone in the waning phase.

This Triple Goddess archetype, present in Celtic, Greek, Hindu, and Native traditions, mirrored a woman's life journey.

Thus, lunar worship became womb worship — and fertility was at its heart.

**Fertility Rites and Full Moon Rituals**

Many ancient societies celebrated fertility festivals in sync with the full moon or spring equinox — a time of flowering, reproduction, and abundance.

### Key Practices:

• Dancing under full moons to invoke fertility blessings.

• Sexual rituals believed to synchronize with moon power and cosmic balance.

• Use of herbs, moonstone, or lunar water to support conception.

• Worship of moon goddesses like:

- Selene/Luna (Greek/Roman)
- Chandra (Hindu)
- Ix Chel (Maya)
- Artemis (protector of women and childbirth)

These rites were often conducted in sacred groves, temples, or moonlit hills.

### Archaeological Clues

• Venus figurines may have symbolized lunar-fertility connection — wide hips, pregnant bellies, exaggerated feminine features.

• Lunar calendar markings on bones and cave walls suggest tracking of menstrual or ovulation cycles.

• Stone circles may have hosted fertility rites tied to lunar standstills or eclipses.

These artifacts indicate a time when life, fertility, and the Moon were inseparable.

### Mysticism and Moon Magic

Even into medieval times, lunar timing was used in:

• Herbal remedies for conception: Gathered during specific phases.

• Astrological birth charts: Moon sign dictated emotional fertility and maternal energy.

• Fertility amulets and moonstone jewelry for enhanced conception.

The new moon was often for setting fertility intentions, while the full moon was for celebrating pregnancies and feminine strength.

### A Global Belief

From tribes in Africa to Vedic India, from Aztec midwives to Norse lunar lore — the Moon's influence on human birth cycles was nearly universally acknowledged.

It wasn't superstition. It was observational science, wrapped in myth and ritual.

### Final Thoughts: Life Beneath Lunar Light

"The Moon doesn't just shape oceans. It shapes life itself."

Even today, studies shows how sleep, hormone levels, even birth rates may subtly align with lunar phases.

The ancients didn't have labs or data models — but they had intuition, observation, and a deep respect for the sky.

To them, the Moon was mother, guide, and gatekeeper of life's great mystery.

And perhaps… it still is.

**Venus figurines**

# 8-Full Moon And Ritualistic Practices

*"Under the full moon, humans danced not just with each other—but with the cosmos."*

### 1. The Eternal Pull of the Full Moon

Since time immemorial, the full moon has captivated the human psyche. Glowing like a celestial lantern, it has stood as a symbol of illumination, revelation, and spiritual climax.

From the jungles of Africa to the high Himalayas, full moon nights inspired fear, awe, and reverence. It was the cosmic moment when darkness bowed before the light.

### 2. Why the Full Moon Matters

#### i. A Symbol of Completion and Release

The full moon represents the peak of the lunar cycle—a moment of culmination. Just as the Moon is at its brightest, our intentions, emotions, and energies are believed to be at their fullest.

In ancient belief systems, this was the time to release what no longer serves—anger, attachments, pain.

In modern psychology and metaphysical practices, the full moon is often used as a checkpoint for emotional detox.

#### ii. Biological & Psychological Impacts

Some studies suggest the Moon affects sleep patterns, menstrual cycles, and even emergency room admissions.

Ancient healers believed the full moon increased fertility, madness, and mysticism—thus the word "lunacy" (from Luna, Latin for moon).

### 3. Rituals Across Time and Cultures

#### a. Ancient Civilizations

Babylonians performed moonlit offerings to Sin, their lunar deity, especially during the full moon.

In Vedic India, full moon days (Purnima) were associated with sacred fasting, meditation, and Chandra Puja.

Native American tribes had names for each full moon like Wolf Moon, Strawberry Moon, and Harvest Moon, each tied to nature's rhythms.

The full moon wasn't just a phase—it was a calendar, a ritual clock, and a spiritual signal.

**b. Eastern Mysticism**

Taoist monks meditated under full moons to absorb "Yin energy".

Buddhist Vesak is celebrated during the May full moon, marking Buddha's birth, enlightenment, and death.

In Japan, Tsukimi is the Moon Viewing Festival—celebrating harvest and moon beauty.

**c. Pagan and Wiccan Traditions**

The full moon is the most sacred time in Wicca. Known as the Esbat, rituals include: Cleansing crystals , Charging sacred tools, Casting spells of forgiveness or manifestation, Moon bathing in silver light.

Bonfires, drumming circles, and herbal offerings are common ways to "draw down the moon"—a ceremony to channel lunar goddess energy.

**4. Common Full Moon Ritual Practices Today**

| Practice | Purpose |
|---|---|
| **Candle Lighting** | Symbolize inner light and intention |
| **Burn Letters** | Emotional release and closure |
| **Herbal Baths** | Energy cleansing and calm |
| **Crystal Charging** | Energetic recharging under moonlight |
| **Moon Journaling** | Self-reflection and gratitude review |

"What you release during the full moon… frees space for what the universe is trying to give you."

## 5. Folklore and Symbolism

In many cultures, wolves howling at the moon symbolize unfulfilled desires or spiritual longing.

Werewolves and shape-shifters were said to gain power on full moon nights—representing the human shadow self.

In folk magic, full moons were times to:Cast love spells, Brew moon elixirs, Seek prophetic dreams.

In art and poetry, the full moon is often portrayed as the "eye of the goddess", watching over earthly cycles.

## 6. Final Thoughts: The Moon as Mirror

The full moon is a mirror of the soul—revealing what we hide, amplifying what we feel. It calls us to remember: we are made of tides and starlight, and every full moon reminds us darkness has a luminous counterpart.

**Babylonians moonlit offerings**

# 09-The Moon Goddess Archetype

*"Before the Moon was science, it was a goddess—watching, whispering, and weaving fate."*

## 1. Introduction: Moon as the Divine Feminine

Long before telescopes measured her craters, humanity personified the Moon as a she—a glowing feminine deity, ever-changing, yet eternal.

Across nearly all ancient cultures, the Moon has been imagined as a divine woman—a goddess of cycles, emotions, fertility, protection, and magic. Her silver glow reflected not just light, but mystery, nurturing, and transformation.

The Moon Goddess isn't one figure—she is a universal archetype, seen in different names, faces, and myths, but always embodying the sacred feminine.

## 2. Global Manifestations of the Moon Goddess

### Selene / Luna – Greece and Rome

- Selene (Greek) or Luna (Roman) rode across the sky in a chariot pulled by white horses.
- Her myth was not of war or conquest—but gentle watchfulness.
- In many temples, she was invoked for protection during childbirth and emotional healing.

### Hecate – Greece

- The dark, mysterious side of the Moon.
- Associated with witchcraft, the underworld, crossroads, and the waning moon.
- Rituals for Hecate were often performed at night, especially during the dark moon.

### Chandra – India

- In Vedic tradition, Chandra is a lunar god, but the Moon's essence is feminine—embodied by goddesses like Parvati, Durga, or Gauri.
- Full moon nights (Purnima) are sacred in Hinduism, observed with fasting, rituals, and moon worship.
- Karwa Chauth, Sharad Purnima, and Vat Purnima show how the Moon is invoked for love, marital harmony, and prosperity.

Hecate - the ancient Greek goddess of witchcraft, magic, the night, crossroads, and ghosts.

**Tsukuyomi – Japan**

- The Moon god in Shinto lore, but like many Eastern traditions, lunar energy is still treated as Yin (feminine).
- The goddess Amaterasu (sun) and Tsukuyomi (moon) reflect cosmic balance of duality—light and dark, active and passive.

**Mama Quilla – Inca Civilization**

- The Incan moon goddess, protector of women.
- Her silver tears were said to fall as moonlight on the Andes.
- She governed menstruation, calendars, and sacred timing.

**Coyolxauhqui – Aztec Mythology**

- A fierce lunar goddess who was dismembered and cast into the sky—a powerful myth representing the Moon's phases.
- Seen as a protector and avenger, she embodied the warrior-moon archetype.

### 3. Themes in the Moon Goddess Archetype

**Fertility and Cycles**

- Just as the Moon waxes and wanes, so do menstrual and agricultural cycles.
- Many ancient rites asked the Moon Goddess to bless wombs, crops, and tides.

**Emotion and Intuition**

- The Moon rules the night and water—both metaphors for emotion, memory, and subconscious.
- She is invoked during rituals of grief, forgiveness, and inner transformation.

**Magic and Mystery**

- The Moon Goddess is associated with prophecy, dreamwork, and divination.
- Witches, shamans, and priestesses across cultures perform rites timed with moon phases, especially the full and new moons.

### 4. Triple Goddess: Maiden, Mother, Crone

In Wicca and neo-pagan traditions, the Moon is honored through the Triple Goddess—representing life's three major stages:

- Maiden – Waxing Moon – youth, curiosity, new beginnings
- Mother – Full Moon – fertility, creation, nurturing
- Crone – Waning Moon – wisdom, endings, transformation

Each phase honors a different face of womanhood, and thus a different stage of lunar energy.

### 5. Artistic and Cultural Depictions

- Statues of Artemis, paintings of Selene, moon masks of Africa, and mandalas in India have all been created to honor her.
- The Moon goddess is often shown with: Crescent crowns, Silver jewelry, Owls, wolves, or serpents as spirit animals ,Moons on her forehead or behind her.

In every symbol, she whispers of balance—between light and shadow, chaos and calm.

### 6. Why This Archetype Still Matters

Even in today's rational, scientific world, the Moon Goddess remains powerful because she reminds us:

- To slow down and feel
- To honor our cycles, not shame them
- To trust intuition over logic sometimes
- That softness is strength, and silence is wisdom

She is a mirror for every soul navigating darkness… and finding their own glow.

### Final Thoughts: Calling the Goddess Within

Every full moon, some still:

- Leave a glass of water in her light
- Light white candles to speak to her
- Dance barefoot under the stars
- Whisper to the night sky…

Not because they believe in mythology—but because they feel something ancient stir within.

The Moon Goddess may be a myth.

But what she reflects—intuition, power, emotion, and grace—is undeniably real.

Coyolxauhqui - the Aztec moon goddess

Selene

MAMA QUILLA

Artemis - the Greek goddess of the hunt, wilderness, wild animals, the moon, and chastity.

# 10 - Moon Symbolism In Tribal Lore

*"Before the Moon was mapped or measured, it lived in fireside stories—wild, sacred, and full of soul."*

## 1. Introduction: The Lunar Spirit in the Tribal Heart

Long before calendars and space missions, the Moon was the primal sky companion of tribal cultures. It appeared in dreams, songs, carvings, dances, and ritual fires. For ancient tribes across continents, the Moon wasn't just a celestial object—it was a living spirit, a watcher in the night sky, a shape-shifter, and a keeper of secrets.

To them, the Moon told stories—not in science, but in symbolism, emotion, and nature's rhythm.

## 2. Across Continents: The Tribal Moon

### North American Tribes

Ojibwe & Cree: Viewed the Moon as a Grandmother—a guardian of women and fertility.

Hopi: Tracked lunar cycles for agriculture and ceremonies.

Inuit: Told stories of moon chases, where the Moon (Aningan) chases the Sun (Malina), creating the cycle of day and night.

"She watches us from above. When wolves howl, it is not to the stars—they howl to her."

### African Tribes

San (Bushmen): Told myths of the Moon dying and being reborn—explaining its phases as cycles of death and resurrection.

Zulu: Believed the Moon had power over fertility and controlled rains and harvest.

Dogon Tribe (Mali): Had mysterious lunar knowledge, associating the Moon with dualism and balance.

**Australian Aboriginals**

Stories of the Moon as a man, wounded and dying repeatedly—explaining the waxing and waning phases.

Believed Moonlight was sacred energy, and Moonrise was a signal to begin nocturnal rituals.

**South American Tribes**

Mapuche (Chile): Saw the Moon as a female spirit, battling the Sun for sky supremacy.

Tupi tribes (Brazil): Used the Moon to time harvest festivals and spiritual fasting.

"When the Moon was full, we painted our bodies, told origin stories, and danced until dawn."

### 3. Key Symbols and Interpretations

| Symbol/Sign | Tribal Meaning |
|---|---|
| **Full Moon** | Fertility, dreams, abundance, peak power |
| **Waning Moon** | Decline, spiritual retreat, letting go |
| **New Moon** | Rebirth, silence, potential |
| **Howling Wolf** | Spiritual connection, ancestral messages |
| **Moonlit Dance** | Tribal unity, transformation, calling spirits |
| **Moon Carvings** | Memory-keeping, sacred timelines |

Moon was not seen in isolation. It was tied to land, animals, menstruation, hunting success, and tribal health.

### 4. Sacred Practices and Rituals

Moon-watching fires were lit during full moons to honor spirits.

Women's moon huts in some tribes offered communal menstrual rest tied to the lunar cycle.

Bone flutes and chants were used to call lunar spirits.

Painting the face with ash or white clay under moonlight was believed to open channels to ancestors.

### 5. Cave Art and Moon Symbols

Some of the earliest cave paintings, especially in France, Spain, and India, depict lunar crescents and dots, possibly representing moon phases.

Petroglyphs from Native American and Aboriginal tribes include moon calendars, animal tracks, and lunar eclipses.

These were not primitive marks—they were calendars, prayers, and poems etched in stone.

### 6. Why This Matters Today

In modern life, we rely on apps and atomic clocks, but we have lost touch with natural time.

The tribal Moon reminds us that:

Time is a circle, not a line.

Emotions are tides, not disorders.

Ritual is not superstition, but memory passed down by the body.

***"The Moon speaks. The tribes listened. And some still do."***

### Final Thoughts: The Wild Moon

Before science, there was story.

Before telescopes, there was instinct.

And before explanations, there was wonder.

The tribal lore of the Moon is a treasure of memory—how humans once connected with the sky using only bones, breath, and belief.

Let us remember the wild Moon… and maybe, listen to her again.

# PART III
# ANCIENT CIVILIZATIONS AND THE MOON

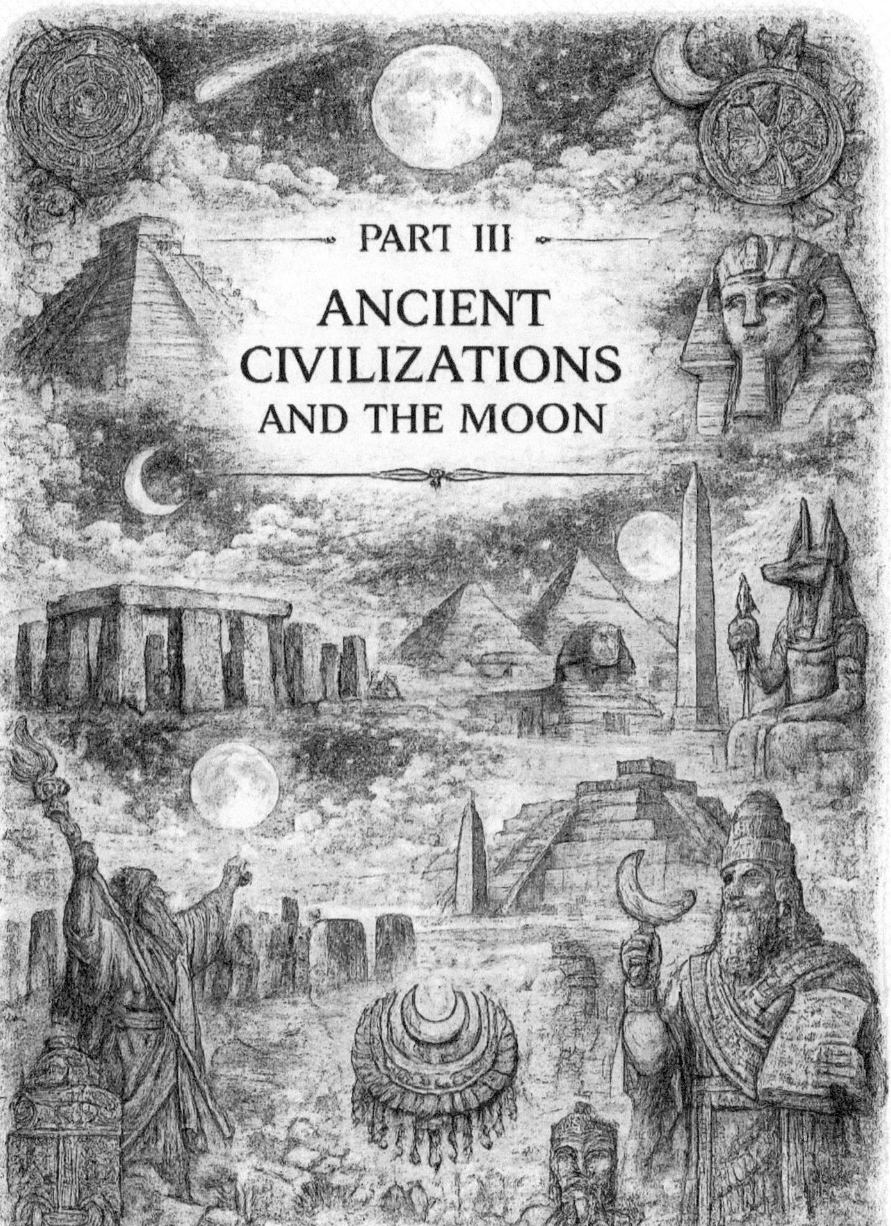

# 11 - Indian Vedic And Puranic Moon Gods – Chandra, Soma

*"He rides a chariot drawn by ten white horses. His light bathes the earth in coolness. His name is Soma… and he is more than just the Moon."*

## 1. The Divine Moon in Indian Thought

In Indian culture, the Moon is not merely a celestial body—it is a deity, an emotion, a cycle, and a symbol of consciousness. From the earliest Ṛigvedic hymns to later Purāṇic texts, the Moon appears as:

- **Chandra:** the gentle god of the Moon
- **Soma:** the elixir of immortality, later personified as the Moon
- **A timekeeper, fertility symbol,** and **astrological influencer**

Let's explore these manifestations—through scriptures, stories, and symbolism.

## 2. Soma in the Vedas: The Elixir and the God

**Soma:** The Plant and the Ritual

In the **Ṛgveda,** Soma originally refers to **a sacred plant and its juice,** believed to grant **divine energy, immortality, and ecstasy.**

**ऋग्वेद9.113.1**

**सोमःपवस्वदेववीतःसदामन्दानःसुवानोवृषा।**

*"O Soma, flowing for the gods, always exhilarated and pure, the mighty bull..."*

Soma was extracted, filtered, and offered in yajñas (rituals), consumed by gods like Indra and Agni, often before battle.

**ऋग्वेद8.48.3**

अपा॑म॒ सोम॑म॒मृता॑ अभू॒माग॑न्म॒ ज्योति॒रवि॑दाम दे॒वान् ।

किं नू॒नम॒स्मान्कृ॑णव॒दरा॑तिः किमु॑ धू॒र्तिर॑मृत॒ मर्त्य॑स्य ॥

*"We have drunk the Soma; we have become immortal; we have attained the light, we have known the gods. What can hostility do to us now, and what, O Immortal, the malice of a mortal?"*

### Soma as the Moon

In later Vedic thought (Brahmana texts), Soma became associated with the Moon, believed to reside in the Moon and wax or wane as it is consumed by gods in heaven.

Thus, each lunar month, Soma is born (new moon), reaches fullness (Purnima), then diminishes (waning moon).

## 3. Chandra in the Purāṇas: The Personified Moon God

### Origin Story

According to Brahmāṇḍa Purāṇa and Viṣṇu Purāṇa, Chandra was born from the ocean of milk during the Samudra Manthan (churning of the ocean). He emerged shining bright, a jewel of the gods.

He is described as:

- Beautiful, white, and youthful
- Mounted on a chariot with ten white horses or an antelope
- Wearing a crescent moon on his forehead

### श्रीमद्भागवतम्5.22.10

*"The Moon, who is also called Soma, is the life of all living entities. He causes the growth of food and strength, and is the source of health."*

### Marriage and Curses

Chandra married the 27 daughters of Dakṣa, representing the 27 Nakṣatras (lunar constellations). But he favored Rohiṇī the most.

Dakṣa cursed Chandra with declining brightness (waning). Appeased later, the curse was modified—Chandra would wax and wane, giving rise to the lunar phases.

### 4. Chandra in Astrology (Jyotiṣa)

In Vedic astrology, Chandra is one of the Navagraha (nine major celestial bodies). He governs:

- Mind (manas), emotions, sensitivity
- Mother, nurturing, home
- Zodiac sign: Cancer (Karka)
- Nakṣatra lord of: Rohiṇī, Hasta, Shravaṇa

**Moon chart (Chandra Lagna)** is central in Indian horoscopes. Full Moon (Purnima) is auspicious, Amāvasyā (New Moon) is linked with ancestral rituals (Pitṛ Tarpan).

### 5. Moon and Agriculture

Traditional Indian farming follows the lunar calendar. Full Moons mark sowing, harvesting, and festivals like:

- **Guru Purnima, Raksha Bandhan, Sharad Purnima**
- Ritual nights of **Mahalaya, Navaratri,** and **Kojagiri**

Moon is also associated with growth of medicinal herbs, rainfall cycles, and Ayurvedic timings.

### 6. Moon in Yoga and Tantric Symbolism

- Idā Nāḍi (left energy channel in the spine) is associated with the Moon—calm, cooling, introspective.
- Soma chakra (subtle energy center above Ajna) is said to secrete amrit during deep meditation.

- In Kundalini Yoga, the Moon balances Sun (Pingala) and enhances intuitive faculties.

## 7. Symbolism and Mantras

- Crescent Moon: immortality, time, waxing/waning of the mind
- White color: purity, coolness, sattva
- Water: Chandra controls all liquids, emotional tides

**Common Chandra Mantra:ॐश्रांश्रींश्रौंसःचन्द्रायनमः**

*"Om Shraam Shreem Shraum Sah Chandrāya Namah"*

Used for calming the mind, balancing emotions, and improving lunar planetary strength.

## 8. Final Reflections: Chandra-Soma – The Eternal Witness

From sacrificial fire to celestial dome, the Indian moon deity has traveled far. In Soma, we find divine intoxication and spiritual nectar. In Chandra, we find the cosmic mirror of our inner world.

Even today, millions perform Chandra Darshan, Karva Chauth, fasts on Mondays, and chant lunar mantras—seeking peace, healing, and balance.

**The Moon in India isn't just up above…**

It flows in rivers of milk, rests on Shiva's forehead, glows in the tears of lovers, and whispers in Vedic hymns

Chandra, the Moon God

# 12 - Chinese Lunar Calendar And The Chang'e Myth

*"Where time is carved in moonlight and immortality floats toward the heavens..."*

## 1. Introduction: A Civilization in Sync with the Moon

From planting crops to celebrating festivals, traditional Chinese life revolved around the Moon. More than a celestial object, the Moon in Chinese culture was a **calendar, goddess, storyteller**, and **symbol of eternal cycles.**

Even today, the Chinese Lunar Calendar, also called the **Yin Calendar,** continues to guide major holidays like Chinese New Year and the **Mid-Autumn Festival**—both deeply lunar in nature.

## 2. The Chinese Lunar Calendar: Time by the Moon

The traditional Chinese calendar is lunisolar, meaning it combines:

Lunar months (based on Moon phases), Solar terms (24 seasonal divisions for farming)

**Key Concepts:**

- Each lunar month begins on a new moon.
- A year normally has 12 lunar months, but a leap month is added every 2–3 years to stay in sync with the solar year.
- Important moon phases: Full Moon (mid-month) = celebrations.

Example: Mid-Autumn Festival always falls on the **15th day of the 8th lunar month**—a full moon night.

This calendar governed everything from agriculture and family planning to astrology and ceremony timing.

## 3. Chang'e: The Immortal Moon Goddess

No story embodies the spirit of the Moon in Chinese culture more than that of Chang'e (嫦 娥 ).

**The Legend:**

Long ago, ten suns rose into the sky, scorching the Earth. The archer Hou Yi shot down nine, saving the world. As a reward, he received the **elixir of immortality.**

But Hou Yi didn't drink it—he wanted to stay with his wife, **Chang'e.**

One day, a treacherous apprentice tried to steal the elixir. To protect it, Chang'e drank it herself… and floated to the Moon, where she became **immortal**.

Now, she lives alone on the Moon, sometimes seen silhouetted during full moon nights.

### 4. The Moon Rabbit: Chang'e's Companion

According to Chinese myth, a jade rabbit (玉 兔 ) lives on the Moon with Chang'e, constantly pounding herbs with a mortar and pestle—perhaps to prepare the elixir again.

This imagery is so powerful that even today, Chinese people refer to the "Moon Rabbit" when looking at the full Moon's markings.

### 5. The Mid-Autumn Festival: A Lunar Celebration

The **Mid-Autumn Festival(中 秋 节)** is China's second-most important festival after Lunar New Year.

**Key Symbols:**

- **Mooncakes:** Round pastries filled with lotus seed paste or egg yolk
- **Family reunions:** Symbolizing wholeness and harmony
- **Lanterns:** Lit to guide good fortune and honor Chang'e

It's celebrated on the 15th day of the 8th lunar month, a night of full moon and reflection.

People gather outdoors to gaze at the Moon, whisper wishes, and admire her glow.

### 6. Astrology and the Moon in Chinese Tradition

Chinese astrology uses the Moon to:

- Determine zodiac compatibility
- Time auspicious days for weddings, moving homes, or business launches
- Predict emotional or energetic shifts (similar to Western Moon sign astrology).

Each zodiac sign's luck can vary depending on monthly Moon phases and lunar transits.

### 7. Symbolism of the Moon in Chinese Philosophy

In Taoism and Confucian thought, the Moon represents:

- Yin: Feminine, passive, introspective energy
- Balance: With the Sun (Yang), creating harmony
- Cycles: Of life, death, and renewal

The Moon is also seen as a mirror of the human soul: ever-changing, yet constant.

### 8. Artistic Depictions

Traditional Chinese paintings, poetry, and music frequently feature the Moon:

**Famous Works:**

- "Quiet Night Thought" by poet Li Bai:
  "Before my bed, the moonlight glows…"
- Scrolls and silks showing Chang'e with flowing robes under a silver sky
- Modern stamps and space missions still bear her image

### 9. The Moon in Modern Chinese Space Missions

China named its lunar missions after Chang'e, blending science with mythology.

- Chang'e 3: First Chinese lander on the Moon
- Chang'e 5: Brought back lunar samples in 2020
- Chang'e 6: In 2024 achieved a historic first by collecting and returning the very first samples from the Moon's mysterious far side.
- Future: Plans for Moon bases and joint missions with Russia

This shows the Moon is not just myth—but a frontier for the future.

### 10. Final Reflection: A Moon That Remembers

The Moon in Chinese culture isn't merely a satellite. It is:

- ✓ A calendar
- ✓ A goddess
- ✓ A memory of love and loss
- ✓ A symbol of unity, balance, and immortality

When a Chinese family eats mooncakes or lights lanterns, they're not just celebrating…

They're reconnecting with ancestors, myth, and the cosmos itself.

# 13 - Egyptian Moon Deities – Thoth And Khonsu

*"In the land where pyramids pierce the heavens and gods walk among men, the Moon was both a guide of wisdom and a master of time."*

## 1. The Moon in Ancient Egypt

The ancient Egyptians were among the first civilizations to track the movements of celestial bodies and assign divine meaning to them. While the Sun God Ra dominated the heavens by day, the Moon ruled the mysteries of night. And unlike some cultures that saw only one lunar deity, Egypt had two major Moon gods—each reflecting a different facet of the Moon's power:

**Thoth** – The god of wisdom, writing, time, and the Moon

**Khonsu** – The youthful lunar god of healing and protection

Both gods played crucial roles in mythology, medicine, ritual, and royal ideology.

## 2. Thoth: The Lunar God of Wisdom and Time

### Origin and Identity

Thoth (Djehuty in Egyptian) was often depicted with the head of an ibis, or sometimes as a baboon, sacred animals linked with the Moon and intelligence.

He wasn't just a moon god—he was a cosmic accountant, tracking time, measuring celestial cycles, and inventing writing (hieroglyphs).

*"He who measures the heavens, brings order to chaos."*

### Thoth and the Lunar Cycle

The phases of the Moon were seen as Thoth's calendar. He used these lunar rhythms to:

- Organize the 12-month Egyptian calendar
- Guide the timing of festivals and rituals
- Balance cosmic dualities like light and dark, life and death

**Patron of Scribes and Magic**

Thoth was credited with:

- Creating language and writing
- Inventing math and astronomy
- Holding the Book of Thoth, containing secrets of the universe

Priests, astrologers, and healers worshipped him as the divine intellect that helped humanity master knowledge.

## 3. Khonsu: The Youthful Healer and Moon Wanderer

### Name and Form

Khonsu (meaning "The Traveler" or "Wanderer") reflects the Moon's nightly journey across the sky.

He is often shown as a young man with a sidelock of youth, wearing a lunar disk and crescent atop his head, sometimes mummified, connecting him with the mysteries of death and rebirth.

### ⚕ God of Healing and Timekeeping

Khonsu was revered for his power to:

- Heal the sick and protect travelers
- Control time, especially the new and full moon cycles
- Ward off evil spirits and nightmares

His cult center in Thebes (Luxor) was part of the great Karnak Temple, where massive lunar rituals were performed.

## 4. Myths and Stories

### The Birth of the Calendar (Thoth's Trick)

In Egyptian myth, the goddess Nut was forbidden from giving birth on any day of the year. Thoth gambled with the Moon, won extra light, and created five additional days—known as the epagomenal days—allowing Nut to give birth to major gods like Osiris and Isis.

This tale links Thoth directly with lunar time and creation.

**Khonsu and the Demon**

One myth tells how Khonsu defeated a demon tormenting a king by using his healing powers and celestial timing. The demon fled under the power of the full moon—a symbol of restoration and purity.

### 5. Lunar Temples and Worship

Karnak Temple Complex in Thebes had a sanctuary for Khonsu, where full moon rituals, chants, and offerings were made.

Hermopolis, the city of Thoth, was a center of lunar studies and sacred writing.

Priests of Thoth were timekeepers, philosophers, and astronomers, managing calendars and conducting precise moon-based rituals.

### 6. Symbolism and Legacy

In Egyptian culture, the Moon stood for:

- Knowledge and Order (Thoth)
- Healing and Protection (Khonsu)
- Time and Cycles
- Judgment and Memory – The Moon was believed to record all deeds, influencing one's fate in the afterlife.

Later cultures like Greek (Hermes) and Roman (Mercury) adopted Thoth's traits.

Even today, the Moon is associated with:

Memory, intuition, and dreams—all echoes of Thoth and Khonsu.

## 7. Final Reflection: Two Faces of the Moon

The ancient Egyptians gave us not one but two lunar mirrors—Thoth, the wise elder, and Khonsu, the youthful healer.

Together, they show how the Moon was more than a distant rock—it was a living presence that:

- Structured civilization
- Protected the weak
- Inspired knowledge and the sacred arts

When we look at the Moon from our modern world, we still feel that ancient pull—a whisper from the Nile under silver light.

**Thoth – the god of wisdom, writing, time and the Moon**

**Khonsu – The youthful lunar god of healing and protection**

# 14 - Mayan Lunar Astronomy And Ritual Sacrifices

*"In the lush jungles of Mesoamerica, where pyramids scrape the skies and jaguars prowl beneath sacred trees, the Moon wasn't just a light in the night — it was a divine timekeeper, a spirit messenger, and a force worthy of awe and blood."*

## 1. The Moon and the Maya

The Maya civilization, flourishing between 2000 BCE and 1500 CE in what is now Mexico, Guatemala, Belize, and Honduras, were among the most advanced astronomers of the ancient world.

They didn't just observe the heavens—they wove celestial rhythms into the very fabric of life, from farming to festivals, birth to burial. And at the heart of it all was the Moon.

## 2. Lunar Calendars and Timekeeping

The Maya maintained multiple calendars, but two in particular reflected lunar influence:

**The Tzolk'in** – Sacred Ritual Calendar

260-day calendar combining 13 numbers and 20 day names

Likely inspired by the gestation cycle of humans and the Moon's role in fertility

**The Lunar Series**

Recorded in Maya codices and monuments, this tracked the 29.5-day lunar cycle ,Included records of:

- New and full moons
- Lunar eclipses
- Moon age calculations

Their astronomers could predict lunar eclipses with astonishing precision—without telescopes!

### 3. The Moon Goddess: Ix Chel

The Maya associated the Moon with Ix Chel, the goddess of:

- The Moon
- Fertility and childbirth
- Weaving and midwifery
- Rain and floods

Often depicted as an old woman with a serpent headdress or a young maiden with a rabbit, she symbolized the changing phases of the Moon.

**The Moon Rabbit**

Just like in Chinese and Aztec cultures, the Maya saw a rabbit in the Moon — not a "man in the Moon."

This rabbit was often associated with fertility and abundance, reinforcing the lunar link with female cycles and renewal.

### 4. Rituals and Sacrifices Aligned with the Moon

Maya religious life was deeply ritualistic, and lunar phases determined the timing of offerings and sacrifices.

**Types of Rituals:**

- Bloodletting by nobles during full moons to commune with gods
- Animal sacrifices timed to lunar eclipses
- Human sacrifices, particularly during important celestial alignments, like a lunar eclipse or new moon

Moon eclipses were often seen as dangerous omens, requiring prayers, noise-making rituals, and offerings to protect the community.

### 5. Eclipses and the Underworld

The Maya believed that during a lunar eclipse, the Moon was being devoured by the lords of Xibalba — the underworld.

To ward off this spiritual attack:

- People would bang drums, shout, or shoot arrows into the sky
- Priests would chant and offer copal incense
- Babies and pregnant women were hidden indoors to protect them

These beliefs weren't just superstition—they were codified in astronomy books, including the Dresden Codex, one of the few surviving Maya manuscripts.

### 6. Lunar Temples and Architecture

Some Mayan temples and pyramids were deliberately built to align with:

- Lunar standstills
- Full moon risings
- Eclipse events

At sites like Uxmal, Copán, and Tikal, staircases and doorways were designed so that moonlight would shine through them at specific times of the year, highlighting divine geometry.

### 7. Legacy of Lunar Wisdom

Though many of their books were destroyed during colonial conquest, the Maya legacy lives on:

Modern Maya communities still honor the Moon in agriculture, childbirth, and folklore

Astronomers study Maya lunar calculations for their mathematical brilliance

The Dresden Codex remains a priceless window into prehistoric lunar astronomy

### Final Thoughts: A Civilization in Sync with the Moon

The Maya didn't see the Moon as distant or passive.

They believed it breathed with the Earth, shaped destiny, and brought balance between life and death, light and shadow.

Their rituals may seem harsh to modern eyes, but in their world, the Moon demanded reverence—and sacrifice.

Every full moon, a Maya priest would climb the steps of a temple, lift eyes to the sky, and whisper:

"Ix Chel, bring us rain, bring us life. We offer blood, so you may offer bounty."

The Caracol at Chichen Itza is an observatory

- **Image courtesy:** *Wikimedia Commons.* **License:** CC BY-SA 2.5

**Types of Rituals**

# 15 - Mesopotamian & Babylonian Lunar Divinations

*"Where the stars spoke and the Moon whispered secrets into the ears of kings."*

## 1. Cradle of Civilization and Cosmic Wisdom

Mesopotamia—literally "land between the rivers"—was the fertile basin between the Tigris and Euphrates. Around 3000 BCE, the Sumerians, Akkadians, Babylonians, and Assyrians looked up to the heavens not just in awe, but in search of divine messages.

To these early cultures, the Moon wasn't just a celestial body. It was a messenger of the gods, a calendar keeper, and most importantly, a prophetic oracle.

## 2. Sin / Nanna – The Mesopotamian Moon God

The Moon god was known by two primary names: Nanna (in Sumerian), Sin (in Akkadian and Babylonian). He was often depicted as a bearded man on a crescent boat sailing through the night sky. His major temples were:

- Ur (E-gish-shir-gal) – A ziggurat devoted to Nanna
- Harran – A city sacred to Sin for centuries

Nanna-Sin ruled over: Time and calendars, Night travel and dreams, Fertility and livestock cycles.

## 3. Lunar Omens and Divination

Mesopotamians believed that every lunar phase, eclipse, or anomaly held meaning.

Their priests, called bārû or diviners, would:

- Record every full moon, crescent, and eclipse
- Consult extensive omen tablets, like the Enuma Anu Enlil series
- Interpret meanings such as:

- "If the Moon is bright but suddenly darkens — war approaches."
- "If a lunar eclipse occurs on the 14th day — the king's health is in danger."

These interpretations guided everything: War or peace, Royal marriages,

Temple offering , Agricultural planning .

### 4. Eclipses: Portents of Doom

A lunar eclipse was often seen as a direct assault on the divine order.

To mitigate the perceived danger:

- A substitute king might be installed temporarily to absorb the curse.
- Rituals including chanting incense, and symbolic sacrifices would be performed.
- The real king might go into hiding for the eclipse's duration.

### 5. Lunar Calendars and Timekeeping

The Babylonian calendar was lunar-solar, with each month beginning at the first visible crescent Moon.

Important features:

- 29- and 30-day months alternating
- Intercalary months added to align with the solar year
- Festivals like Akitu (New Year) aligned with lunar phases
-

### 6. The Legacy of Lunar Knowledge

Babylonian lunar astronomy influenced:

- Jewish calendar systems
- Greek and Hellenistic lunar science
- Islamic calendar traditions later on

**Modern scientists still study:**

- Clay tablets containing lunar ephemerides (e.g., the MUL.APIN)
- Observational methods of Mesopotamian priest-astronomers

### Final Thoughts: The Moon as Divine Clock and Oracle

To Mesopotamians and Babylonians, the Moon was a living calendar, a divine adviser, and a celestial witness.

When a priest stood atop a ziggurat under a crescent sky, tablet in hand, watching the stars move across the heavens—he wasn't just stargazing. He was listening to gods speak in light and shadow.

In many ways, modern lunar science walks in the footsteps of these ancient seers.

**Ziggurat of Ur**

PART IV
MOON AND
THE CALENDAR
MEASURING TIME

# 16 - Lunar Vs Solar Calendars

# The Evolution Of Timekeeping

*"Time is not just marked by clocks, but by shadows of the Moon and the paths of the Sun."*

## 1. Origins of Timekeeping: Watching the Sky

Before clocks and calendars, humans turned their eyes upward—to the sky.

- The **Moon**, with its visibly changing shape over ~29.5 days, became nature's first calendar.
- The **Sun**, with its steady daily rise and fall, became a symbol of continuity, seasons, and solar years.

Thus, two parallel timekeeping systems arose:

- The **Lunar calendar:** based on the Moon's phases.
- The **Solar calendar**: based on Earth's orbit around the Sun.

## 2. Lunar Calendars – Rhythms of the Night Sky

**Structure:**

- A lunar month = ~29.5 days
- 12 lunar months = ~354 days (11 days shorter than a solar year)

**Key Characteristics:**

- Months begin with the new moon or first crescent
- Dates drift across seasons unless corrected
- Intercalary months (leap months) are added in some cultures to sync with seasons

**Civilizations that followed lunar or lunisolar calendars:**

- Babylonian and Hebrew calendars
- Chinese lunar-solar calendar
- Islamic Hijri calendar (strictly lunar, no intercalation)
- Indian Panchang (luni-solar)
- Mayan lunar cycles for rituals

### 3. ☀️Solar Calendars – The Cycle of the Seasons

**Structure:**

- Based on the tropical solar year (~365.24 days)
- Tracks Earth's position around the Sun
- Divides time by solstices, equinoxes, and the seasons

**Examples:**

- Ancient Egyptian calendar: among the first to track solar years (~365 days)
- Gregorian calendar: modern global standard (leap year system)
- Julian calendar: precursor to Gregorian

Solar calendars became essential for:

- Agriculture and harvests
- Predicting seasonal weather patterns
- Organizing large-scale societies

### 4. Luni-Solar Calendars – Bridging the Two Worlds

Some civilizations developed hybrid systems, combining both Moon and Sun:

- They followed lunar months but adjusted periodically to match solar years.
- Used intercalary months to keep festivals and seasons aligned.

Notable Luni-Solar Systems:

- Indian Hindu Calendar (Panchang)
- Chinese Calendar (used for festivals like Lunar New Year)
- Hebrew Calendar (used for Jewish rituals)

These calendars ensured that seasonal festivals remained in the right part of the year, while still honoring lunar traditions.

### 5. The Challenges of Lunar Calendars

- Drifting seasons: Without correction, lunar months move backward through the year (e.g., Ramadan rotates across seasons every 33 years).
- Intercalation confusion: Adding extra months wasn't always systematic or consistent.
- International alignment: As trade and empires grew, diverse calendars created synchronization problems.

### 6. Cultural and Religious Impacts

The Moon became deeply tied to religion, ritual, and spiritual timing:

- Ramadan, Eid, Passover, Chinese New Year, and Mahashivratri all rely on lunar dates.
- The Sun, meanwhile, symbolized harvest, power, and order.
- Ancient temples and monuments like Stonehenge, Egyptian obelisks, and Chichen Itza are aligned with solar events.

### 7. Scientific Advancements and Standardization

The transition from lunar to solar (or standardized) calendars was fueled by:

- Better astronomical tools (e.g., astrolabes, sundials)
- Desire for agricultural precision
- Need for international and administrative unity

The Gregorian calendar (introduced in 1582 by Pope Gregory XIII) became the gold standard for civil use—correcting drift in the earlier Julian calendar.

### 8. Why It Still Matters Today

Many traditional societies still observe lunar cycles for rituals, farming, and navigation.

Modern calendars like the Chinese, Hindu, Hebrew, and Islamic remain active.

Astronomy apps and lunar observatories continue to track the Moon with high accuracy.

Even today, many people's biological rhythms, menstrual cycles, and cultural behaviors are subtly influenced by the Moon's phases.

### Final Thoughts: A Dance of Time

The story of calendars is not just math—it's the story of how humans interpret the sky, mark memory, and celebrate life.

The Moon teaches us rhythm.

The Sun teaches us reason.

And together, they gave rise to calendars—the very foundation of civilization.

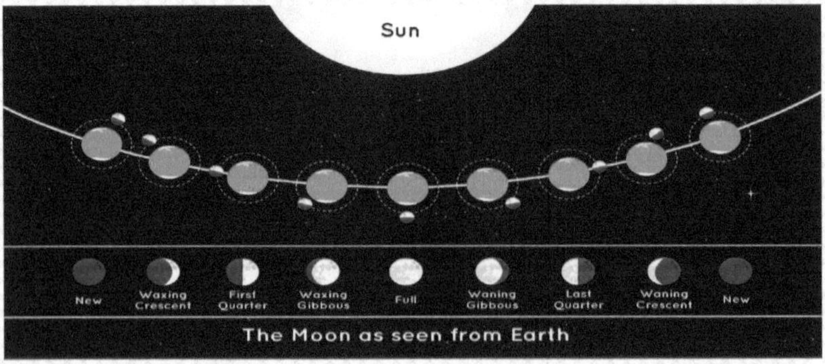


Moon Phases: NASA Science

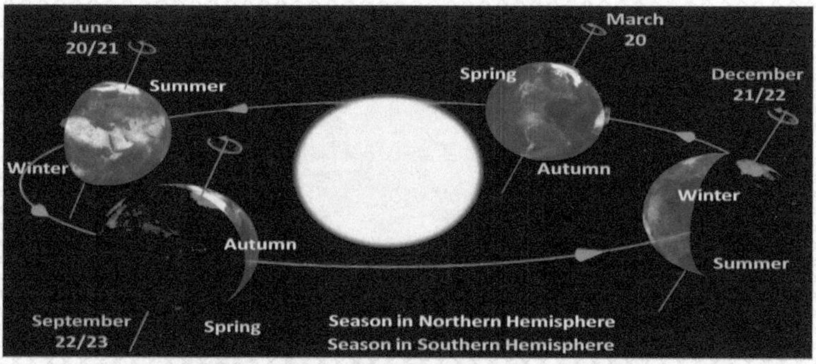


Earth's Orbit Around Sun :CC0

**Egyptian Calender :** Kom Ombo Temple Calendar 2.

Aztec calendar stone. By El Comandante – CC BY-SA 3.0 (Wikimedia Commons)

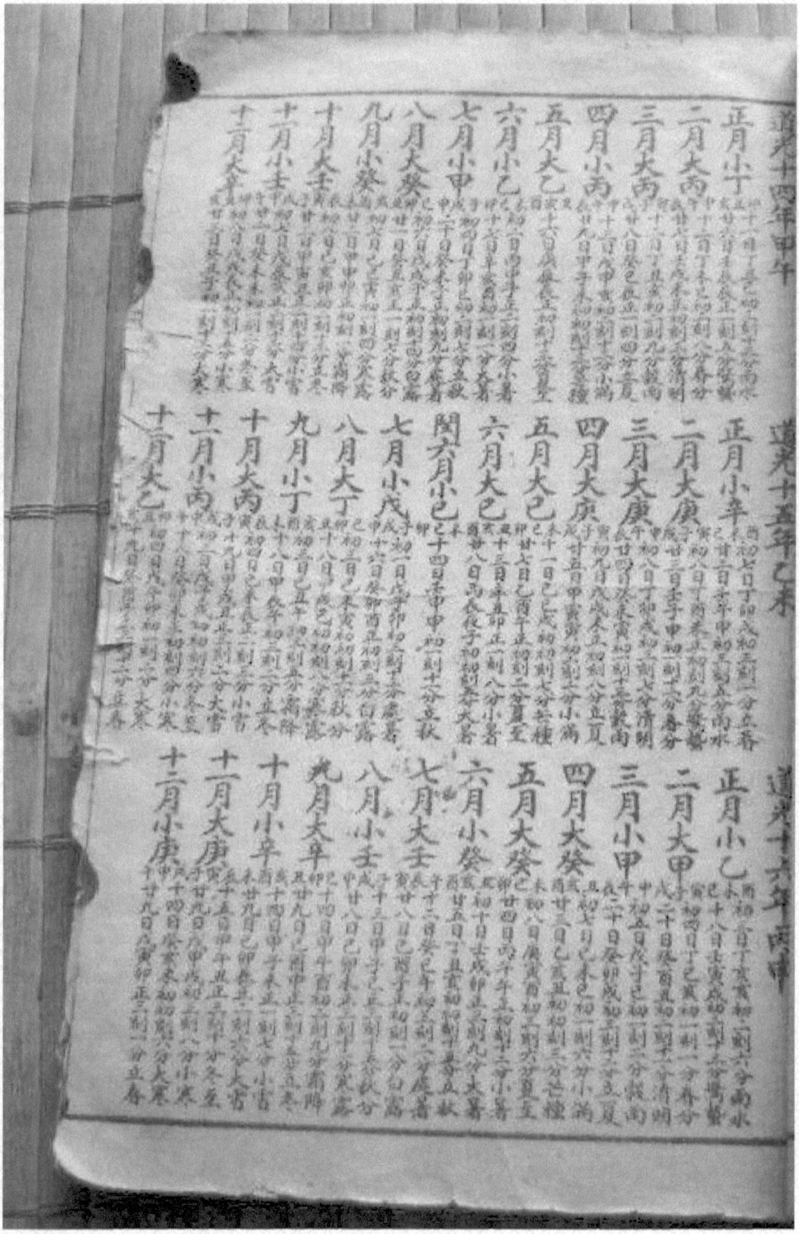

Chinese lunisolar calendar for 1834, 1835, and 1836 By Orienomesh-wCC BY-SA 4.0

Depiction of the 19 years of the Metonic cycle as a wheel, with the Julian date of the Easter New Moon, from a 9th-century computistic manuscript made in St. Emmeram's Abbey (Clm 14456, fol. 71r)

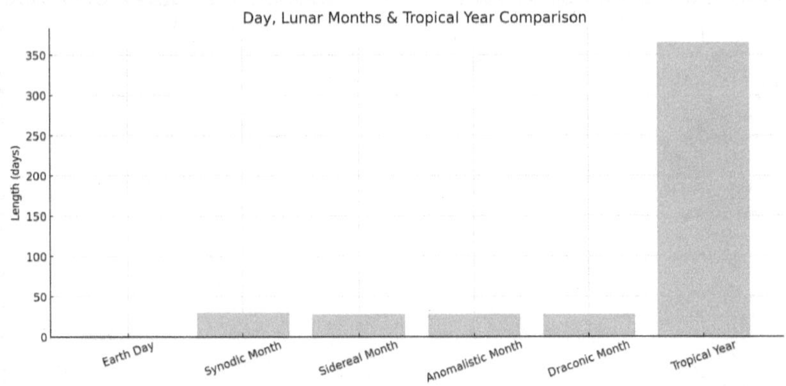


• Earth Day (1 day) • Synodic Month (29.5306 days)

• Sidereal Month (27.3217 days) • Anomalistic Month (27.5546 days)

• Draconic Month (27.2122 days) • Tropical Year (365.2422 days)

# 17 - The Islamic Hijri Calendar

## A Lunar Devotion To Time

*"Verily, the number of months with Allah is twelve months..." — Surah At-Tawbah (9:36)*

The Islamic Hijri Calendar, is a purely lunar system used predominantly in the Islamic world. It is not just a method of timekeeping—it's deeply intertwined with religion, ritual, and spiritual memory, beginning with the Prophet Muhammad's migration (Hijrah) from Mecca to Medina in 622 CE.

### 1. Structure of the Hijri Calendar

- 12 lunar months, each 29 or 30 days
- A Hijri year totals about 354 or 355 days, which is approximately 11 days shorter than a solar year.
- This difference causes Islamic months to shift through the solar year, rotating through all seasons roughly every 33 solar years.

**The 12 Months of the Islamic Calendar:**

| No. | Month Name | Arabic Name | Meaning / Importance |
|---|---|---|---|
| **1** | Muharram | ٱلْمُحَرَّم | Sacred month |
| **2** | Safar | صَفَر | Departure |
| **3** | Rabi' al-Awwal | رَبِيع ٱلْأَوَّل | Month associated with the birth of Prophet ﷺ |
| **4** | Rabi' al-Thani | رَبِيع ٱلثَّانِي | Second spring |
| **5** | Jumada al-Awwal | جُمَادَىٰ ٱلْأُولَىٰ | Dryness begins |
| **6** | Jumada al-Thani | جُمَادَىٰ ٱلثَّانِيَة | Second dry month |
| **7** | Rajab | رَجَب | One of the sacred months |
| **8** | Sha'ban | شَعْبَان | Preparation month |
| **9** | Ramadan | رَمَضَان | Month of fasting |
| **10** | Shawwal | شَوَّال | Eid al-Fitr follows |

| 11 | Dhul-Qi'dah | ذُو ٱلْقَعْدَة | Month of rest |
|---|---|---|---|
| 12 | Dhul-Hijjah | ذُو ٱلْحِجَّة | Month of Hajj |

## 2. Religious Significance

The Hijri calendar is fundamental to Islamic rituals:

- Ramadan – Month of fasting and revelation of the Quran
- Eid al-Fitr – Celebrated after Ramadan ends
- Dhul-Hijjah – Month of Hajj and Eid al-Adha
- Ashura (10th Muharram) – Commemoration of events in Islamic history

All these are observed based on the moon's appearance, highlighting the lunar devotion of Islam.

## 3. Moon Sighting: A Ritual in Itself

Unlike pre-calculated calendars, the Hijri calendar often depends on visual sighting of the new moon (Hilal), especially for beginning Ramadan and Eid.

This has led to:

- Different countries starting festivals on different days
- Increased use of telescopes and lunar observatories in modern Islamic societies
- Integration with apps to predict Hilal visibility with astronomical accuracy

Still, spiritual tradition often favors naked-eye observation, preserving the ancient rhythm between sky and soul.

## 4. Theological Philosophy

Islam's use of the lunar calendar reflects deeper ideas:

- Humility and surrender to natural signs
- Detachment from rigid time structures—symbolizing divine control over human planning
- Encouragement to observe the sky, fostering celestial awareness in every believer

### 5. Global and Cultural Presence

The Hijri calendar:

- Is official in Saudi Arabia, used in Islamic legal contexts
- Is parallel to the Gregorian calendar in many Muslim-majority nations
- Is used on mosque calendars, religious apps, and Islamic event posters
- Still marks birthdays of saints, Islamic New Year, and Sufi celebrations

### 6. Science and the Moon in Islam

In early Islamic civilization:

- Astronomers like Al-Battani and Al-Zarqali improved lunar calculations
- Observatories tracked lunar phases centuries before telescopes
- The Moon is mentioned over 20 times in the Quran, often symbolizing guidance, measurement, and divine signs

"It is He who made the Sun to be a shining glory and the Moon to be a light..." – Surah Yunus (10:5)

### 7. Final Thoughts: A Moon-Lit Faith

The Hijri calendar is a testament to a civilization that looked to the sky not for conquest, but for reflection.

In a world run by solar schedules, Islam's lunar path is a reminder that time can be sacred, cyclical, and beautifully connected to nature.

When the crescent moon is sighted, a billion hearts prepare to fast, to pray, to celebrate. This is not just a calendar—it is a cosmic conversation.

# 18 - Hindu Tithi System - Amavasya And Purnima

*"The Moon is not merely a light in the night sky in Hinduism—it is the keeper of time, emotion, and karma."*

## The Tithi System

In Hindu timekeeping, Tithi refers to a lunar day, a key element in the Panchang (Hindu calendar). Unlike Western days that follow the Sun, Hindu days follow the phases of the Moon.

There are 30 Tithis in a lunar month, divided into two fortnights:

**Shukla Paksha** (waxing phase) – New Moon (Amavasya) to Full Moon (Purnima)

**Krishna Paksha** (waning phase) – Full Moon to New Moon

Each Tithi is around 12 degrees of angular distance between the Sun and Moon, approximately 0.95 solar days long, meaning some Tithis may start or end mid-day.

## Amavasya – The New Moon Night

Amavasya is the night of the New Moon, when the Moon is not visible. It's a time traditionally associated with:

- Introspection and detachment.
- Ancestral worship (Pitru Tarpan and Shraddha)
- Occult rituals and tantra

* Many observe Amavasya with fasting and meditation. In Mahalaya Amavasya, Hindus perform rituals for their ancestors to bless the living family.

In Hindu scriptures, Amavasya is a powerful time where divine and ghostly energies are active. It's also considered ideal for spiritual seekers and sadhakas.

### Purnima – The Full Moon

Purnima is the night of the full Moon, when it appears fully lit. Spiritually, this is a time of:

- Completion and spiritual abundance
- Festivals and celebrations
- Worship of Satya, Vishnu, Lakshmi, and other deities

Guru Purnima, Buddha Purnima, and Sharad Purnima are some of the most sacred observances.

From a symbolic perspective, Purnima reflects the culmination of inner growth. The mind is said to be most restless but also most receptive during this time.

### Symbolism: Amavasya vs Purnima

| Feature | Amavasya | Purnima |
|---|---|---|
| **Moon visibility** | Invisible | Fully visible |
| **Energy type** | Inward, mystical, dark | Radiant, celebratory, expansive |
| **Activities** | Meditation, Shraddha, austerity | Worship, charity, learning |
| **Mood of mind** | Quiet, reflective | Emotional, elevated |
| **Ideal for** | Letting go, spiritual cleansing | Receiving blessings, growth |

**Psychological and Spiritual Influence**

According to yogic and tantric texts:

The Moon influences Chitta (mind-stuff) and manas (conscious mind).

**Amavasya** is the best time to break patterns, go inward, or do shadow work.

**Purnima** amplifies devotion (bhakti), creativity, and intuitive insight.

The phases of the Moon are believed to directly affect the nadi (energy channels) and doshas (bodily energies) in Ayurveda and Yoga.

**Scriptural References**

Rigveda (1.164.48) refers to the Moon's phases as measures of time and seeds of rhythm.

The Chandogya Upanishad mentions the Moon as the abode of the ancestors.

The Bhagavad Gita (8.24–26) describes the lunar cycles as part of the spiritual exit paths of the soul.

**The Cosmic Clock**

Through Amavasya and Purnima, Hindus didn't just track time—they lived in alignment with cosmic forces.

Rituals were timed to lunar psychology.

Fasting, donations, pilgrimages, and mantras were calibrated to the Moon's mood.

It was an ancient form of celestial mindfulness—far before modern science began to understand circadian rhythms or lunar effects on hormones.

**Modern-Day Relevance**

**Even today:**

Indian festivals like Karva Chauth, Maha Shivaratri, Holi, and Raksha Bandhan are synced with Tithis.

Astrology and muhurat (auspicious timing) depend heavily on Tithi.

Mobile apps like Drik Panchang offer Tithi tracking, preserving ancient knowledge.

**Conclusion**

The Moon doesn't just govern tides—it governs minds, rituals, and dharma in Hindu tradition. Amavasya and Purnima are not simply days—they're energetic portals embedded in time.

Understanding the Tithi system gives us not just cultural knowledge, but a way to live rhythmically, in harmony with the Moon, the cosmos, and ourselves.

| Tithi Name | Meaning | Paksha | Lunar Angle (Sun-Moon) |
|---|---|---|---|
| **Pratipada** | 1st day | Shukla/Krishna | 0°-12° / 180°-192° |
| **Dwitiya** | 2nd day | Shukla/Krishna | 12°-24° / 192°-204° |
| **Tritiya** | 3rd day | Shukla/Krishna | 24°-36° / 204°-216° |
| **Chaturthi** | 4th day | Shukla/Krishna | 36°-48° / 216°-228° |
| **Panchami** | 5th day | Shukla/Krishna | 48°-60° / 228°-240° |
| **Shashthi** | 6th day | Shukla/Krishna | 60°-72° / 240°-252° |
| **Saptami** | 7th day | Shukla/Krishna | 72°-84° / 252°-264° |
| **Ashtami** | 8th day | Shukla/Krishna | 84°-96° / 264°-276° |
| **Navami** | 9th day | Shukla/Krishna | 96°-108° / 276°-288° |
| **Dashami** | 10th day | Shukla/Krishna | 108°-120° / 288°-300° |
| **Ekadashi** | 11th day | Shukla/Krishna | 120°-132° / 300°-312° |
| **Dwadashi** | 12th day | Shukla/Krishna | 132°-144° / 312°-324° |
| **Trayodashi** | 13th day | Shukla/Krishna | 144°-156° / 324°-336° |
| **Chaturdashi** | 14th day | Shukla/Krishna | 156°-168° / 336°-348° |
| **Purnima/ Amavasya** | Full/New | Purnima/Amavasya | 180° / 360° or 0° |

# 19- Jewish And Chinese Lunar Months

## Sacred Rhythms Of Time

***From ancient scrolls in Jerusalem to jade temples in Beijing, two great civilizations have tracked time not by the blazing Sun—but by the gentle phases of the Moon. The Jewish and Chinese lunar calendars offer rich traditions in how lunar months shaped festivals, rituals, and even national identity.***

### 1. Jewish Lunar Calendar – A Sacred Sync

The Hebrew calendar is lunisolar, meaning it follows the Moon for months and Sun for seasons. Months begin with the new Moon, and years are adjusted with leap months (Adar II) to sync with the solar cycle.

**Scriptural Significance:**

- Genesis 1:14 – "Let there be lights in the sky to separate the day from the night... for signs, festivals, days, and years."

- The Torah commands that festivals like Passover must occur in spring, hence leap years ensure proper seasonal alignment.

**The 12 (Sometimes 13) Months:**

| Month | Typical Gregorian Range | Notable Festivals |
|---|---|---|
| Nisan | March–April | Passover (Pesach) |
| Iyar | April–May | Lag BaOmer |
| Sivan | May–June | Shavuot |
| Tammuz | June–July | 17th of Tammuz |
| Av | July–August | Tisha B'Av |
| Elul | August–September | Reflection Month |
| Tishrei | Sept–Oct | Rosh Hashanah, Yom Kippur, Sukkot |
| Cheshvan | October–November | Quiet month |

| | | |
|---|---|---|
| Kislev | November–December | Hanukkah |
| Tevet | December–January | Fasts |
| Shevat | January–February | Tu BiShvat (Tree Festival) |
| Adar | February–March | Purim |
| Adar II | (Leap years only) | Second Purim |

**Metonic Cycle:** Every 19 years, 7 leap years (with 13 months) are added to sync lunar months with solar years.

### 2. Chinese Lunar Calendar – Harmony with Heaven

The Chinese calendar is also lunisolar. Each lunar month starts with the new Moon and culminates with the full Moon. It has been used for thousands of years to plan agriculture, festivals, and astrology.

**Origin & Legacy:**

- Legend attributes the calendar to the Yellow Emperor Huangdi (2698 BCE).

- Modern Chinese people still celebrate festivals based on the lunar calendar, even though the Gregorian calendar is used in civic life.

### Major Lunar Festivals:

| Festival | Month | Description |
|---|---|---|
| **Chinese New Year** | 1st lunar month | Most important festival, marked by firecrackers, lion dances, and red envelopes. |
| **Lantern Festival** | 15th day of 1st month | Full Moon celebration with lanterns and riddles. |
| **Dragon Boat Festival** | 5th month | In memory of poet Qu Yuan, marked by boat races and sticky rice dumplings. |

| | | |
|---|---|---|
| **Mid-Autumn Festival** | 8th month (Full Moon) | Mooncakes are eaten and families reunite under the bright Moon. |
| **Ghost Festival** | 7th lunar month | Spirits are believed to roam; offerings are made. |

Zodiac Connection: Each lunar year is associated with one of the 12 zodiac animals and tied to Taoist cosmology, yin-yang theory, and the Five Elements.

**Comparison:**

| Aspect | Jewish Calendar | Chinese Calendar |
|---|---|---|
| Type | Lunisolar | Lunisolar |
| Year Start | Nisan (Religious), Tishrei (Civil) | Varies (Chinese New Year) |
| Religious Basis | Torah, Talmud | Taoism, Confucianism, folklore |
| Adjustments | Leap month 7 of 19 years | Leap month 7 of 19 years |
| Major Festivities | Passover, Sukkot, Yom Kippur | Lunar New Year, Mid-Autumn |
| New Moon Significance | Rosh Chodesh (new month) | Each month starts with New Moon |

**Mystical Symbolism:**

- In **Jewish Kabbalah,** the Moon represents the Shekinah (divine feminine).

- In **Daoist mysticism,** the Moon symbolizes yin energy and reflection.

Why This Matters: These calendars show deep respect for nature and timekeeping—guiding when to fast, feast, plant, or pray.

# Chapter 20: Buddhist Full Moon Festivals

## The Sacred Lunar Light

***"As the moon brightens the darkest night, so does mindfulness illuminate the soul."– Buddhist Saying***

### The Moon in Buddhism

In Buddhism, the full moon is not merely an astronomical event. It represents awakening, reflection, and the rhythm of life. Many of Buddhism's most sacred festivals are tied to full moon days, known as Uposatha. Across different Buddhist traditions — Theravāda, Mahayāna, and Vajrayāna — the full moon symbolizes purity, enlightenment, and the ever-changing nature of existence.

**What is Uposatha?**

Uposatha (Pali; Sanskrit: Upavasatha) is the Buddhist equivalent of a holy day, observed on new moon, full moon, and quarter moon days.

Originating from ancient Indian customs and adopted by early Buddhists, these lunar observances are meant for purification, reflection, and renewed commitment to the path of Dhamma.

On these days, laypeople observe the Eight Precepts, while monks gather to chant the Patimokkha (rules of discipline).

### Major Full Moon Festivals in Buddhism

**1. Vesak (Buddha Purnima / Visakha Bucha)**

Date: Full moon of Vesakha (April-May)

Observed in: Sri Lanka, India, Thailand, Myanmar, Cambodia, Laos, Nepal

Significance:Birth, Enlightenment, and Parinirvana (death) of Gautama Buddha

Most important Buddhist festival

Traditions:Temple visits, lanterns, alms-giving, sermons

Bathing Buddha statues, acts of compassion

**2. Magha Puja (Sangha Day)**

Date: Full moon of Magha (February-March)

Observed in: Thailand, Cambodia, Laos, Myanmar

Significance: Commemorates a spontaneous gathering of 1,250 enlightened monks to hear Buddha's teaching

Symbol: The Ovada Patimokkha – Buddha's core teachings on peace and restraint

Traditions: Candlelit processions, sermons, meditation

**3. Asalha Puja (Dhamma Day)**

Date: Full moon of Asalha (July)

Significance: Buddha's first sermon after enlightenment – the Dhammacakkappavattana Sutta ("Turning the Wheel of Dhamma")

Traditions: Dhamma talks, practice of generosity, meditation retreats

**4. Pavarana and End of Vassa**

Date: Full moon of Ashvini (October)

Significance: Marks the end of the 3-month Vassa (monk's rains retreat)

Pavarana: monks invite mutual feedback and atonement

Symbolism:

Renewal, reflection, and inner cleansing

**Ritual Practices on Full Moon Days**

Observing Eight Precepts (celibacy, fasting, right speech, etc.)

Temple visits and meditation

Offering flowers, incense, and lamps

Releasing birds or animals as a karmic act of compassion

Listening to Dhamma talks and chanting

## Symbolism of the Full Moon in Buddhist Thought

| Moon Symbolism | Buddhist Interpretation |
|---|---|
| • Full Moon | Enlightenment, completion, inner clarity |
| • New Moon | Renewal, introspection, beginner's mind |
| • Waxing Moon | Growth in virtue, developing insight |
| • Waning Moon | Letting go, renunciation, simplification |

The full moon is often linked to mindfulness, clarity of Samadhi (meditative concentration), and Vipassana (insight). Just as clouds hide the moon, mental defilements hide inner truth. Meditation, then, is the act of "clearing the clouds."

## Artistic and Literary Expressions

Thangka paintings in Tibetan Buddhism often depict full moons behind bodhisattvas

Zen Haikus use the moon as a metaphor for awakening:

"The moon's reflection / in still water — the mind / when truly quiet."

## Conclusion: A Global Lunar Light

From Bodh Gaya to Bangkok, the full moon still governs the spiritual calendars of millions of Buddhists. It calls for silence in chaos, discipline in freedom, and mindfulness amid distraction.

The moon is not just above us — in Buddhist thought, it's within us. Ever-changing, yet always pre

Buddha's first sermon after enlightenment

# PART V

# MOON IN ASTROLOGY, SPIRITUALITY, AND MYSTICISM

# 21- Vedic Astrology And Moon's Nakshatras

*– The Lunar Highways of Karma and Consciousness*

## Jyotisha – The Eye of the Vedas

In ancient India, astrology wasn't superstition—it was science fused with spirituality. The Moon, in Vedic astrology, holds a far more personal role than in Western systems. Known as Chandra, the Moon governs emotion, mind, memory, motherhood, and the subconscious.

The Moon's journey across the sky was mapped with a unique system of 27 lunar mansions, or Nakshatras—a celestial code said to shape our destiny, psychology, and karmic path.

## Why the Moon, Not the Sun?

While Western astrology focuses on solar signs, Vedic astrology is lunar-centric. That's because:

The Moon moves quickly — it spends just 1 day in each Nakshatra, making it a better clock for tracking human emotion and time-bound rituals.

In Vedic thought, the mind (Manas) is the seat of karma. And the Moon is its ruler.

Hence, one's Janma Nakshatra (birth star) is based on where the Moon was placed at the time of birth—not the Sun.

## The 27 Nakshatras: Lunar Constellation Portals

Here are the 27 Nakshatras, each spanning 13°20' of the zodiac:

| No. | Nakshatra | Symbol | Deity | Key Traits / Themes |
|---|---|---|---|---|
| 1 | Ashwini | Horse's Head | Ashwini Kumars | Quick healing, pioneering, beginnings |

| | | | | |
|---|---|---|---|---|
| 2 | Bharani | Yoni (Womb) | Yama (God of Death) | Transformation, discipline, sexuality |
| 3 | Krittika | Blade / Flame | Agni (Fire God) | Sharpness, cleansing, courage |
| 4 | Rohini | Chariot / Red Cow | Brahma | Charm, fertility, beauty, creativity |
| 5 | Mrigashira | Deer's Head | Soma (Moon God) | Curiosity, seeker, romance |
| 6 | Ardra | Teardrop | Rudra (Storm God) | Change, emotions, destruction, storm |
| 7 | Punarvasu | Bow / Quiver | Aditi (Mother of Gods) | Renewal, return, optimism |
| 8 | Pushya | Cow's Udder | Brihaspati (Jupiter) | Nourishment, spirituality, generosity |
| 9 | Ashlesha | Serpent | Nagas (Serpent Spirits) | Mysticism, cunning, entanglement |
| 10 | Magha | Throne / Royal Seat | Pitris (Ancestors) | Heritage, royalty, pride, tradition |
| 11 | Purva Phalguni | Swing / Bed | Bhaga (Sun God) | Pleasure, love, luxury |
| 12 | Uttara Phalguni | Hammock / Bedpost | Aryaman (Companion God) | Duty, friendship, legacy |
| 13 | Hasta | Hand | Savitar (Sun God) | Dexterity, skill, communication |
| 14 | Chitra | Pearl / Gemstone | Tvashtar (Cosmic | Aesthetics, beauty, brilliance |

| | | | Architect) | |
|---|---|---|---|---|
| 15 | Swati | Coral / Sword | Vayu (Wind God) | Independence, flexibility, movement |
| 16 | Vishakha | Triumphal Arch | Indra & Agni | Determination, ambition, duality |
| 17 | Anuradha | Lotus | Mitra (Friendship) | Loyalty, devotion, collaboration |
| 18 | Jyeshtha | Amulet / Earring | Indra (King of Gods) | Seniority, leadership, charisma |
| 19 | Mula | Root | Nirriti (Destruction) | Deep research, letting go, root cause |
| 20 | Purva Ashadha | Elephant Tusk | Apah (Water God) | Victory, cleansing, belief |
| 21 | Uttara Ashadha | Plank / Bed | Vishwadevas (Universal Gods) | Law, truth, endurance |
| 22 | Shravana | Ear / Listening | Vishnu | Learning, fame, communication |
| 23 | Dhanishta | Drum / Musical Tone | Vasus (Elemental Gods) | Rhythm, wealth, celebration |
| 24 | Shatabhisha | 100 Physicians | Varuna (Cosmic Order) | Healing, secrets, solitude |
| 25 | Purva Bhadrapada | Twin Legs / Sword | Aja Ekapada (Fiery Deity) | Transformation, intensity |
| 26 | Uttara Bhadrapada | Serpent / Twins | Ahir Budhnya (Water Serpent) | Patience, mystery, deep wisdom |

| 27 | Revati | Fish / Drum | Pushan (Nourisher) | Protection, journeys, prosperity |
|---|---|---|---|---|

## Birth and Destiny: The Role of Janma Nakshatra

In Vedic astrology, your Janma Nakshatra determines:

- Your mental and emotional makeup
- Strengths and weaknesses
- Compatibility in relationships
- Auspicious rituals (muhurta), naming conventions (Namakarana)
- Dasha system (planetary periods) that govern life phases

Example:If born in Rohini, one may be artistic, sensual, and charming—but also emotionally intense and possessive.

## Nakshatras and Ritual Life

In Hindu rituals, lunar days and Nakshatras guide:

- Festivals (e.g., Guru Purnima on Purnima + Pushya)
- Marriage muhurats
- Upanayan samskaras
- Agricultural activities
- Temple consecrations and Chandi Yagnas

Even Vastu Shastra (architectural timing) often aligns with Nakshatra positions.

## Spiritual Symbolism of the Moon's Orbit

The Moon's waxing and waning mirrors the growth and decline of mind-states

Nakshatras represent spiritual stations—like portals on the soul's journey

Several Nakshatras have links to mythical figures—e.g., Rohini (beloved of Krishna), Pushya (connected with Guru), Shravana (symbol of hearing and learning)

### Conclusion: The Mind is the Moon

In Vedic consciousness, the Moon is not an object—it's the mirror of the soul.

Just as the Moon reflects the Sun's light, the mind reflects the Atman (soul).

To track the Moon is to track our own patterns of desire, emotion, and transformation.

The Nakshatras aren't superstition—they are ancient psychological archetypes, embedded in nature's clock, whispering secrets of karma in the language of stars.

# 22 - Western Astrology

## Moon Signs And Emotions

*"If the Sun is how we shine in the world, the Moon is how we feel when we're alone."*

### The Moon: The Mirror of Emotion

In Western astrology, the Moon holds the key to our inner world. While the Sun sign (your birthday zodiac) represents ego, identity, and outward expression, the Moon sign governs the hidden emotional landscape: how we feel, respond, nurture, and seek safety.

Your Moon sign is determined by the position of the Moon at the exact moment of your birth, and it changes signs roughly every 2.5 days—meaning even twins born hours apart might have different Moon signs and thus emotional temperaments.

### The Moon in the Birth Chart

**In a Western natal chart, the Moon represents:**

- Emotional Needs – What soothes us, how we seek comfort.
- Early Childhood Conditioning – The emotional climate we grew up in.
- Intuition & Reactions – Gut feelings, moods, and instinctive responses.
- Inner Self – Our private identity, distinct from the public persona.

A person with Moon in Cancer may be deeply nurturing and emotionally responsive, while someone with Moon in Aquarius might seem detached but values emotional independence.

### The 12 Moon Signs and Their Emotional Styles

| Moon Sign | Emotional Nature |
|---|---|
| **Aries** | Impulsive, passionate, quick to react, needs freedom. |
| **Taurus** | Calm, sensual, craves stability and comfort. |
| **Gemini** | Restless, curious, processes emotions through talking. |
| **Cancer** | Deeply sensitive, nurturing, easily hurt, craves safety. |
| **Leo** | Warm, expressive, needs recognition and affection. |
| **Virgo** | Anxious, self-critical, seeks order and purpose in emotions. |
| **Libra** | Peace-loving, avoids conflict, craves harmony in relationships. |
| **Scorpio** | Intense, private, emotionally deep, prone to emotional extremes. |
| **Sagittarius** | Optimistic, adventurous, avoids heavy emotions. |
| **Capricorn** | Reserved, disciplined, takes responsibility for feelings. |
| **Aquarius** | Detached, idealistic, expresses emotions intellectually. |
| **Pisces** | Empathic, dreamy, easily overwhelmed, deeply intuitive. |

### Moon Phases and Moods

Just as the Moon waxes and wanes in the sky, Western astrologers believe our moods fluctuate with the lunar phases:

New Moon: Time of introspection, planting intentions.

Waxing Moon: Growth, optimism, action.

Full Moon: Emotional climax, revelations, heightened energy.

Waning Moon: Release, reflection, letting go.

The lunar cycle is used in astrological forecasting, aligning emotional work and decisions with the energy of the Moon.

### Psychology Meets Astrology

Modern psychological astrologers, like Liz Greene and Dane Rudhyar, have merged Jungian concepts with Moon symbolism. The Moon becomes the "Mother archetype", representing not just the biological mother, but emotional nurturing, memory, and unconscious habits.

Emotional trauma, childhood attachment styles, and inner security are often analyzed through the Moon sign, aspects to other planets, and house placement in the natal chart.

### The Moon's Cultural Expansion in the West

In Western history, the Moon has shifted from goddess worship (Selene, Luna, Diana) to scientific curiosity and back again into the mystical with the New Age movement.

The resurgence of interest in Moon rituals, lunar journaling, Moonstone crystals, and lunar yoga practices reflects a modern reconnection with the intuitive, emotional, and spiritual side of human life—long represented by the Moon.

### Final Thoughts

Western astrology paints the Moon not as a distant celestial object, but as a mirror to our emotional soul.

"The Moon knows what the heart tries to hide."

Whether we follow horoscopes or study deep natal charts, understanding the Moon's position in our chart can be one of the most profound steps in emotional self-discovery.

## 23- Moon In Tarot, Alchemy, And Occult Practices

***"The Moon whispers not in words, but in symbols, dreams, and shadows. Those who walk the path of the occult look up at her light not for certainty, but for mystery."***

**The Moon in Tarot**

In the Tarot, the Moon card (Major Arcana XVIII) is one of the most enigmatic and unsettling. Depicted with a glowing moon shining between two towers, a dog and a wolf howling at it, and a crayfish crawling out of the water, it is a card of illusion, dreams, and the unconscious mind.

The Twin Towers: They represent the boundary between the known and unknown worlds.

The Dog and Wolf: Civilization and wilderness, tamed and untamed instincts.

The Crawfish: Emerging from the deep waters of the subconscious.

When drawn upright, the Moon card speaks of intuition, psychic visions, dreams, and hidden truths. In reversed form, it warns of deception, fear, and confusion.

For the mystics, the Moon card is not about clarity—it's about trusting one's inner senses when logic fails.

**⚗The Moon in Alchemy**

Alchemy—the ancient mystical science of transforming matter and spirit—considered the Moon a symbol of silver, the noble metal paired with gold (the Sun).

Silver & Gold: Just as the Sun (gold) represents the soul's perfection, the Moon (silver) embodies the reflective, receptive principle—intuition, imagination, and the unconscious.

The White Work (Albedo): One of the key alchemical stages, where the blackened "matter" (Nigredo) is purified into white. This stage was often symbolized by the Moon's light, cleansing darkness with reflection rather than fire.

Alchemy's Duality: The Moon was always paired with Mercury, the shapeshifting metal, representing transformation, cycles, and hidden wisdom.

For alchemists, the Moon was not just a celestial body but a spiritual force guiding the inner purification of the human soul.

## The Moon in Occult & Esoteric Practices

The Moon has always held sway over the occult—secret traditions that move between religion, magic, and mysticism.

Witchcraft & Lunar Phases: Witches traditionally time spells with the Moon.

1. New Moon: Beginnings, intentions, planting seeds.
2. Waxing Moon: Growth, attraction, abundance.
3. Full Moon: Power, manifestation, divination.
4. Waning Moon: Banishing, releasing, endings.

Moon & Divination: Scrying bowls filled with water, gazed upon under moonlight, were believed to reveal visions. Dreams during the full moon were thought to be prophetic.

Secret Societies: In Hermetic and Rosicrucian traditions, the Moon was a gatekeeper between the physical and astral worlds. Its symbolism appears in Kabbalah, where it is linked to Yesod, the foundation of the mystical Tree of Life—the channel between higher and lower realms.

## Why the Moon Is Central in the Occult

The Moon's power lies not in certainty but in ambiguity. Where the Sun is clarity, the Moon is shadow. Where the Sun is day, the Moon is night. Mystics, alchemists, and occultists have always looked to her light as a mirror of the unconscious, a guide through uncertainty, and a reminder that not all truths shine—they sometimes whisper in dreams.

"To study the Moon is to accept that truth wears masks. And in those masks—whether in Tarot, alchemy, or occult rites—we glimpse not answers, but deeper questions."

**Tarrot Card**

**Alchemist**

# 24-Black Magic, Witchcraft & Full Moon Rituals

## 1. The Moonlit Stage of Mysticism

On nights when the Moon swells into perfect fullness, the world appears subtly altered. Shadows sharpen, silence deepens, and the familiar landscape takes on an uncanny glow. Across civilizations, such nights were believed to exist *between worlds* — neither wholly earthly nor entirely divine.

The full moon was seen as a **liminal gateway**, a threshold where the boundaries between the natural and the supernatural weakened. Ancient observers noticed that tides rose higher, animals behaved strangely, and human emotions intensified. To pre-scientific societies, this could only mean one thing: the Moon was *active*, radiating an invisible power.

Thus, full moon nights became the preferred stage for rituals — prayers for healing, fertility rites, initiations, divination, and, in darker traditions, acts of manipulation and forbidden magic. The Moon did not merely illuminate these rites; it was believed to *witness*, *empower*, and sometimes *judge* them.

In this silver light, human fears, desires, and hidden ambitions were laid bare.

## 2. Witchcraft and the Full Moon

In European folk traditions, the full moon was regarded as a time of **heightened psychic energy**. Witches — healers, midwives, herbalists, and spiritual practitioners — aligned their work with lunar phases, believing the Moon governed growth, decay, intuition, and transformation.

One of the most sacred full-moon practices was **"Drawing Down the Moon."** In this ritual, a priestess symbolically invited the lunar goddess to descend into her, merging human and divine consciousness. This was not possession, but communion — a moment of insight, prophecy, and empowerment.

Contrary to later Church narratives, much of this witchcraft was **protective and life-affirming**:

- Blessing crops and livestock
- Assisting childbirth
- Healing illness through herbs and lunar timing
- Creating charms for love, protection, or guidance

Yet secrecy was essential. Knowledge passed quietly from woman to woman, hidden from authorities who feared what they could not control. Over time, misunderstanding hardened into accusation, and reverence was recast as heresy.

### 3. Black Magic: The Darker Associations

While witchcraft itself was broad and morally complex, **black magic** referred specifically to practices believed to cause harm — manipulation of fate, curses, coercion of spirits, or domination of another's will.

Medieval grimoires meticulously timed spells according to lunar phases. The full moon, believed to magnify intention, was considered the moment when illusions became real and unseen forces could be bent — for good or ill.

Black magic thrived on **secrecy and fear**. Candlelit rituals, blood oaths, and whispered incantations fed public imagination. Whether such acts truly wielded supernatural power or simply psychological terror, their reputation endured.

Even today, folklore warns of:

- Hexes cast under full moons
- Enchanted objects charged by moonlight
- Dreams or visions following nocturnal rituals

The Moon became associated not just with mystery, but with danger — a mirror reflecting humanity's darker impulses.

### 4. Folklore and Fear: From Europe to Asia

In medieval Europe, villagers spoke in hushed tones of witches flying across the moonlit sky. Owls, black cats, and wolves — creatures of the night — became lunar symbols of fear. The Moon was imagined as a silent witness to secret sabbaths held in forests and ruins.

In India, **Purnima** (full moon) carried a dual meaning. For many, it was sacred — a time for fasting, prayer, and devotion. Yet certain tantric traditions performed intense rituals under full moon light, believed to awaken dormant energies. Outsiders, unable to distinguish spiritual discipline from superstition, labeled these rites as dark or dangerous.

Across Southeast Asia, moonlit nights were associated with wandering spirits, restless ancestors, and shapeshifting beings. The Moon did not cause fear — it revealed what was already believed to exist.

## 5. The Dual Nature of Full Moon Rituals

Perhaps the most fascinating aspect of lunar belief is its **duality**.

The same full moon that inspired tales of curses also inspired hymns of love and healing. In Wicca and neo-pagan traditions, **Esbats** — monthly moon rituals — celebrate the full moon as a peak of spiritual clarity.

These rituals focus on:

- Gratitude and release
- Emotional healing
- Manifestation of intentions
- Honoring the feminine divine

Thus, the Moon embodies contradiction: feared and adored, cursed and worshiped, shadowed and luminous. It reflects not morality, but **human intention**.

## 6. Case Studies of Ritual Practices

### European Witches' Sabbaths

Historical records describe gatherings on full-moon nights where participants danced, chanted, and communed with nature spirits. Many such accounts were exaggerated by inquisitors, but archaeological and folkloric evidence suggests communal rituals centered on lunar cycles did exist.

### Tantric Moon Rituals (India)

Certain yogic traditions viewed moonlight as spiritually nourishing. Meditative practices involved gazing at the Moon to calm the mind and balance inner energies. More extreme rites, often misunderstood, contributed to fear-based narratives.

### African & Caribbean Traditions

In Vodou and Santería, the full moon is ideal for ancestral communication. Offerings are made, drums played, and spirits invited — not for harm, but guidance and healing.

### Modern Neo-Paganism

Today's practitioners consciously reject the term "black magic." Full moon rituals emphasize ecological balance, personal empowerment, and psychological insight rather than supernatural domination.

## 7. Symbolism of the Full Moon in Witchcraft & Black Magic

Across cultures, recurring themes emerge:

**Illumination of the Unseen** – Moonlight reveals hidden truths

**Transformation** – Linked to shape-shifting and rebirth myths

**Climax of Power** – The peak of magical cycles

**Ambivalence** – Neither purely good nor evil

The Moon does not choose sides. It amplifies.

## 8. The Cultural Legacy

Even in modern society, the full moon retains its psychological grip. Emergency rooms report heightened anxiety, people complain of sleeplessness, and popular culture continues to portray moonlit nights as uncanny.

From horror films to fantasy novels, the Moon remains the perfect backdrop for magic and mystery. Yet beneath the superstition lies a deeper truth: the Moon has always symbolized the **human subconscious** — reflective, cyclical, emotional, and powerful.

## Conclusion: Light and Shadow

The full moon does not create darkness — it reveals it.

Whether invoked for healing or feared for its power, lunar rituals expose humanity's eternal struggle between control and surrender, reason and intuition, light and shadow.

In witchcraft, black magic, and mysticism, the Moon remains what it has always been:
**a mirror, not a master — reflecting the intentions of those who dare to work beneath its light.**

BLACK MAGIC, WITCHCRAFT
& FULL MOON RITUALS

# 25-Astral Travel And Full Moon Meditations

*"When the world sleeps under the silver gaze of the Moon, seekers claim the spirit awakens to wander the invisible realms."*

## The Lunar Gateway

For millennia, mystics, shamans, and yogis have believed that the Full Moon is not just a celestial event, but a cosmic doorway. Its radiant pull is said to thin the veil between the physical and the astral worlds. Many traditions—ranging from Indian yogic texts to medieval European occult manuscripts—describe the Moon as a mirror between realms, reflecting hidden truths and guiding souls into higher dimensions.

Astral travel, often called out-of-body experience (OBE), is the practice of the soul or consciousness leaving the physical body and journeying across subtle planes of existence. When combined with meditations under the Full Moon, practitioners claim visions sharpen, dreams intensify, and the boundary between the "seen" and "unseen" becomes porous.

## Ancient Roots of Astral Travel

Egyptian Mystics – Ancient Egyptian priests believed the soul could leave the body, traveling with the Moon god Thoth, who guided spirits through cosmic passages.

**Indian Yogic Philosophy** – Yogis described "sukshma sharira" (the subtle body) which could separate from the gross body during deep meditation, aided by mantras under the Full Moon.

**Greek Mystery Schools** – Plato and Pythagoreans wrote of "soul flights" where knowledge was gathered beyond earthly limits.

**Shamanic Traditions** – Tribal shamans worldwide used drumming, chanting, and moonlight rituals to induce astral journeys for healing and prophecy.

## Full Moon Meditation Practices

During a Full Moon, the Earth, Moon, and Sun align in a way that many believe amplifies energy fields. Spiritual practitioners developed specific meditations:

Lunar Gazing (Chandra Trataka): Sitting under moonlight, staring at the Moon to awaken third-eye vision.

Mantra Recitation: Chanting specific mantras (like Om Chandraya Namah) to tune consciousness with lunar vibrations.

Guided Astral Projection: Lying in stillness while visualizing the soul gently floating out of the body, rising above under the Moon's glow.

Dream Work: Keeping a journal after Full Moon meditations, as lucid dreams and symbolic visions become more vivid.

## The Science Behind the Mystery

Modern science, though skeptical of astral travel, acknowledges the Moon's strong impact on circadian rhythms, sleep cycles, and psychological states. Studies show:

Sleep during a Full Moon may be lighter and filled with more REM activity.

People often report heightened creativity, intuition, and vivid dreams.

Meditation under moonlight synchronizes brain waves, inducing trance-like states.

While science may not confirm "soul travel," it does recognize the neurological conditions that make such experiences feel real.

## Case Studies and Cultural Narratives

European Witches: Medieval accounts describe witches gathering under the Full Moon to perform astral flights, aided by chants and herbal potions.

Tibetan Yogis: Through "dream yoga," monks practiced conscious exploration of dream states, akin to astral journeys, especially during lunar rituals.

Modern Practitioners: Contemporary meditation groups hold Full Moon retreats, where group energy is believed to enhance astral experiences.

### Why This Matters

The study of astral travel and Full Moon meditations is not just about mysticism. It reflects humanity's eternal quest to transcend limits, to reach beyond the body, and to connect with the cosmos. Whether literal journeys of the soul or symbolic explorations of the mind, these practices remind us of a universal truth:

**The Moon is more than a rock orbiting Earth.**

It is a spiritual mirror, guiding seekers toward self-discovery, expanded consciousness, and cosmic belonging.

### Closing Thought

*'When you close your eyes under the Full Moon, you don't just meditate—you set sail. The body sleeps, but the soul remembers its wings."*

PART VI
MOON IN FOLKLORE, LITERATURE, AND ART

# 26: Moon In Classical Literature And Poetry

***"Poets may not explain the birth of the Moon, but they have given her countless lives."***

From the earliest written texts to the great masterpieces of world literature, the Moon has always been a muse—mysterious, inspiring, and eternal. Classical literature and poetry transformed the Moon from a cold object in the sky into a symbol of love, longing, melancholy, and transcendence.

## The Moon in Greek and Roman Literature

Homer in The Iliad and The Odyssey often described the Moon's glow as a guide to warriors and wanderers. Its steady presence in the night sky symbolized order in a chaotic world.

Ovid, in Metamorphoses, associated the Moon with the goddess Diana (Artemis), portraying her as both protector and untouchable maiden.

Roman poets like Horace and Virgil frequently invoked Luna in their odes, connecting the lunar cycle with harvests, fertility, and fate.

For the ancients, the Moon was not just scenery—it was a character in the drama of gods and men.

## The Moon in Medieval and Persian Poetry

Dante Alighieri (Divine Comedy) placed the souls of those bound by vows in the sphere of the Moon, symbolizing imperfection and human frailty.

Persian poets like Rumi and Hafez often used the Moon as a metaphor for divine beauty or the beloved. To Rumi, the Moon

was both an unreachable star and the silent witness to love's intoxication.

## The Moon in Shakespeare and Renaissance Drama

Shakespeare adored the Moon. In A Midsummer Night's Dream, the Moon governs the rhythm of love, madness, and transformation.

In Romeo and Juliet, Juliet warns against swearing love "by the inconstant Moon," for it changes every night—symbolizing fickleness.

Elizabethan poets saw the Moon as a mirror of human emotions: joy, betrayal, passion, and sorrow.

### The Romantic Moon

The Romantic era (18th–19th centuries) gave the Moon her most lyrical expressions.

William Wordsworth wrote of the Moon as a companion of the lonely traveler.

Percy Bysshe Shelley in The Cloud and To the Moon portrayed it as a weary yet eternal spirit, reflecting human despair and hope.

John Keats linked the Moon to mortality and beauty, as in Endymion, where the shepherd dreams of eternal union with the goddess of the Moon.

Here, the Moon was no longer divine authority but a sympathetic friend to human hearts.

### Eastern Classical Poetics

In Japanese haiku, poets like Bashō captured the Moon in minimalist strokes: a pond, a frog's leap, and the reflection of the Moon.

In Chinese Tang poetry, Li Bai and Du Fu made the Moon a symbol of longing, especially for distant friends or lovers. Li Bai famously wrote, "We share the same Moon, though we are far apart."

In Indian Sanskrit poetry, the Moon (Chandra) is praised as cool-rayed (śītāṁśu), a comforter of lovers separated by fate. Kalidasa, in Meghaduta, describes the Moon as soothing balm to the sorrowful.

### Why the Moon Matters in Literature

The Moon has always been more than a celestial body—it is:

A mirror of human emotion.

A symbol of time, change, and impermanence.

A bridge between lovers, dreamers, and the divine.

Poetry and literature reveal that while astronomy studies the Moon's matter, imagination makes it immortal.

**Final Thought:**

When we read classical poetry, we see the same Moon our ancestors saw—yet each generation clothes it with its own meanings. The Moon is both constant and ever-changing, much like humanity itself.

# Classical Literary Excerpts on the Moon

## 1. William Shakespeare (England, 16th c.)

From *Romeo and Juliet*, Act 2, Scene 2:
*"O, swear not by the moon, th' inconstant moon,*
*That monthly changes in her circled orb,*
*Lest that thy love prove likewise variable."*

Here, Shakespeare warns against swearing love by the Moon, symbolizing her changeability.

---

## 2. Kālidāsa (India, 4th–5th c. CE)

From *Meghadūta (Cloud Messenger)*, Sanskrit:
*"प्रियतमविरहव्यथा हृदयं दग्धं चन्द्रकिरणैः शीतलं भवति"*
**Translation:** "The heart, scorched by separation from the beloved, becomes cool in the rays of the moon."

Kālidāsa frequently linked the Moon with love, longing, and soothing comfort.

---

## 3. Jalāl ad-Dīn Rūmī (Persia, 13th c.)

*"At night, I open the window*
*and ask the moon to come and press its face against mine.*
*Breathe into me.*
*Close the language-door and open the love-window."*

Rumi's poetry makes the Moon a symbol of divine love and mystical union.

### 4. Homer (Greece, 8th c. BCE)

From *The Iliad* (Book VIII):
*"As when in heaven the stars about the moon*
*Look beautiful, when all the winds are laid,*
*And every height comes out, and jutting peak*
*And valley, and the immeasurable heavens*
*Break open to their highest, and all the stars*
*Shine, and the shepherd gladdens in his heart."*

The Moon becomes a pastoral guide, illuminating shepherds in the night.

---

### 5. Omar Khayyām (Persia, 11th c.)

From *Rubáiyát*:
*"The nightingale cries to the rose, the Moon looks down in disdain.*
*Ah, life is fleeting! Pour the wine, for tomorrow we may not remain."*

The Moon here reminds humans of time's fleeting nature and mortality.

---

### 6. Rabindranath Tagore (India, 19th–20th c.)

From *The Crescent Moon*:
*"The sleeping child smiles like a crescent moon in the cradle of the sky."*

Tagore uses the Moon as a metaphor for innocence and childhood wonder.

---

Together, these excerpts show how **different civilizations—Greek, Sanskritic, Persian, English—used the Moon as metaphor for love, fate, constancy, and divine mystery.**

# 27 : Moon As The Muse In Japanese Haiku

*"Few things in nature have stirred the Japanese poetic soul as deeply as the Moon. In the delicate world of haiku, the Moon is not just an object in the sky—it is a mirror of fleeting beauty, silence, and emotion."*

## Haiku and the Spirit of Impermanence (Mono no Aware)

In Japanese aesthetics, mono no aware refers to the bittersweet awareness of impermanence—the fleeting beauty of life. The Moon, constantly waxing and waning, became a natural symbol of this concept. Just as cherry blossoms fall too soon, the Moon never stays in one phase for long.

For poets, capturing the Moon in a haiku was not only about describing its shape, but also the emotion it evoked—loneliness, longing, or serene beauty.

## Classic Haiku Masters on the Moon

**1. Matsuo Bashō (1644–1694)**

"Clouds now and again

give a soul some respite from

moon-viewing — and love."

Here Bashō uses clouds as a metaphor for love's interruptions—both the Moon and human emotion shine through fleetingly.

**2. Yosa Buson (1716–1784)**

"Harvest moon—

walking around the pond

all night long."

Buson, known for his painterly imagery, captures the Moon not as distant, but as a companion for a sleepless walk.

**3. Kobayashi Issa (1763–1828)**

"The moon tonight—

I climb the mountain path

with the old dog."

Issa's Moon is personal, intimate, and tied to everyday life. Even with an aging companion, the Moon becomes a witness to tenderness and time.

### Tsukimi: Moon Viewing as Cultural Poetry

The tradition of tsukimi (月見, "moon-viewing") dates back to the Heian period (794–1185). Nobles and poets would gather to compose poems, drink sake, and gaze at the autumn Moon. This was not just an aesthetic pastime—it was a spiritual practice of tuning one's heart to the rhythms of nature.

Rice dumplings (tsukimi dango), pampas grass, and clear sake were offered, blending ritual with poetry. For the Japanese, the act of looking at the Moon was itself a kind of verse.

### Haiku as Lunar Snapshots

Haiku, limited to 17 syllables (5-7-5), became the perfect vessel for expressing the instantaneous magic of the Moon. Unlike long epics, a haiku catches the shimmer of moonlight on water, the silence of a snowy night, or the melancholy of a waning crescent—like a photograph carved in words.

### Sample Haiku Collection

"Clear autumn moon—

pond snails are moving about

in the thin water." — Buson

"Looking at the moon

with the morning after eyes—

last night's love fades." — Anonymous Edo poet

"The moon's brightness—

a thousand miles away,

same as at home." — Bashō

### Visual Symbolism of the Moon in Japanese Art

In traditional ukiyo-e woodblock prints, the Moon often appears above temples, mountains, or travelers on lonely paths. These images inspired and echoed haiku, blending visual and literary art into a single lunar language.

### Why It Matters

The Japanese haiku tradition shows us that the Moon is more than a celestial body—it is a bridge between nature and the heart, a constant reminder of impermanence and beauty.

Even today, poets and dreamers around the world echo what the haiku masters of Edo Japan once felt: that in a single glance at the Moon, the universe becomes poetry.

# 28 - Moon In Romantic Art And Music

*"The Moon, pale and eternal, became the soul of longing in Romantic imagination—an emblem of love, sorrow, and beauty that words alone could never capture."*

## The Romantic Movement and the Moon

The Romantic era (late 18th – 19th century) was a reaction against reason and industrialization. Poets, painters, and musicians sought to return to nature, emotion, and the mystical depths of the human spirit. Among all natural symbols, the Moon stood supreme.

For Romantics, the Moon was not just a celestial object. It was a confidant of the lonely, the silent witness of lovers, and the melancholy light of night wanderers.

## The Moon in Romantic Art

1. Caspar David Friedrich (1774–1840)

The German painter often depicted figures gazing at the Moon, backs turned, lost in reverie. His works like "Two Men Contemplating the Moon" (1819) embody the Romantic mood: solitude, friendship, and the vastness of the cosmos.

2. J.M.W. Turner (1775–1851)

Turner's landscapes glowed with dramatic skies. His moonlit seascapes captured both the tranquility and turbulence of nature—an interplay of light and shadow that echoed Romantic passion.

3. Francisco Goya (1746–1828)

Though often dark, Goya's nocturnal scenes used the Moon as a spectral presence. In works like "Witches' Sabbath," the Moon watches over rituals, blending beauty with unease.

## The Moon in Romantic Music

1. Beethoven's "Moonlight Sonata" (1801)

Perhaps the most famous lunar-inspired piece. Though not named by Beethoven himself, the sonata's first movement, with its haunting arpeggios, evokes a moonlit lake—mysterious, tender, and mournful.

2. Schubert's "Serenade" (Ständchen, 1828)

Schubert's lyrical songs often invoked the Moon as a companion of love and longing. The Moon's glow becomes the stage-light of romance.

3. Chopin's Nocturnes (1830s–1840s)

The very word "nocturne" means "night piece." Chopin's piano works shimmer like moonlit gardens, delicate yet filled with unspoken emotion.

4. Debussy's "Clair de Lune" (1890, revised 1905)

Though slightly later, this impressionist masterpiece channels the Romantic soul—dreamlike piano phrases that flow like silver moonbeams on water.

### The Moon as a Symbol of Emotion

Love and Desire: Lovers often swore their devotion under moonlight. The Moon symbolized eternal passion.

Melancholy and Solitude: The waning Moon mirrored fading hope, loss, or yearning.

Mystery and Dream: The Moon was the doorway to imagination, inspiring visions beyond reality.

### Poetic Echoes

"The night walked down the sky with the moon in her hand." — Frederic L. Knowles

"When I behold the skies aloft / Passing the pageantry of dreams, / The cloud whose bosom, cygnet-soft, / A couch for fairy fancy seems— /

Methinks no heart should beat with care / Against so pure a sky as this."
— Thomas Hood

### Visual and Musical Fusion

Romantic salons often combined music, poetry, and art in moonlit gatherings. Poets read verses while pianists played nocturnes, with the Moon as muse to all. The blending of forms reflected the Romantic belief that the Moon was not confined to one medium—it belonged to the universal imagination of humanity.

### Why It Matters

The Moon's role in Romantic art and music shows how a celestial body became an emotional archetype. It was not studied as in science, or worshiped as in myth, but felt—as a presence in the night sky that mirrored the soul's deepest emotions.

Even today, when Beethoven's sonata plays or when we stand before Friedrich's paintings, we sense the same quiet magic: the Moon, eternal, reflecting both light and longing.

**"O, swear not by the moon, the inconstant moon,**

**That monthly changes in her circled orb…"**

**— Shakespeare, Romeo and Juliet**

**"Look, the moon is risen in the silver night,**

**And all the stars are sleeping…"**

**— Kalidasa, Meghaduta**

**"The moon stays bright when it doesn't avoid the night."**

**— Rumi**

MOONLIGHT LOVERS:
A JOURNEY THROUGH TIME

# 29- Western Fairytales – Werewolves & Lunacy

***"The full moon rises, casting silver shadows across forests and villages. In its glow, legends awaken — of beasts that walk like men, and men who howl like beasts."***

**Moonlight and Madness**

For centuries, the Moon has been more than a celestial body to Western imagination. In Europe, particularly during the Middle Ages, the full moon became a symbol of transformation, mystery, and even terror. Unlike the serene moon deities of Asia, or the agricultural lunar calendars of the East, the Western moon often carried darker associations.

**Two key motifs dominate these legends:**

Werewolves – humans who transformed into wolves under the light of the full moon.

Lunacy – the belief that moonlight could disturb the mind, driving people mad.

Both ideas shaped centuries of storytelling, law, and even medicine. Let's explore how these myths evolved, how they were feared, and why they still fascinate us today.

**Werewolves – The Shape-Shifters of the Night**

**Origins of the Wolf Curse**

The earliest whispers of werewolf tales date back to Greek and Roman mythology. The Greek legend of Lycaon, a king punished by Zeus for his cruelty, describes him being transformed into a wolf. The very word "lycanthropy" comes from this myth.

The Romans carried similar tales into their empire. Wolves, both feared and admired, represented the wildness beyond human control. By the Middle Ages, Christianity rebranded the wolf as a creature of the devil, and thus the werewolf became a cursed soul bound to darkness.

### The Full Moon Trigger

In medieval Europe, people believed transformations only occurred under the light of the full moon. This connection was not coincidental:

The moon's cycle regulated time and rituals.

Its brightness pierced through night, allowing nocturnal activity to be visible.

Folklore made the leap: if wolves howled at the moon, then cursed men must answer it too.

Thus, the lunar cycle became a supernatural clock, marking the nights when terror might strike.

### × Werewolf Trials in Europe

Much like the witch hunts, werewolf trials became common between the 15th and 18th centuries. Records from France, Germany, and Switzerland describe villagers accusing neighbors of being "loup-garous" (wolf-men).

In 1521, Gilles Garnier, known as the "Werewolf of Dole," was executed in France after confessing to killing children in wolf form.

Court documents even include "guides" for spotting werewolves:

Excessive hairiness

Eyebrows that met in the middle

Strange behavior during full moons

Most of these trials reflected fear of outsiders or mental illness, but they fed into the enduring legend.

### The Werewolf in Literature & Pop Culture

Medieval ballads often warned against wandering into forests during the full moon.

In 18th–19th century Gothic literature, werewolves became metaphors for repressed human desires.

Modern films like The Wolf Man (1941) or An American Werewolf in London (1981) cemented the full moon as the essential trigger of transformation.

The myth persists today in fantasy novels, TV series (Teen Wolf, True Blood), and games — the moonlit werewolf is as immortal as the creature itself.

## Lunacy – When the Moon Disturbs the Mind

### Etymology and Belief

The very word "lunacy" comes from luna, Latin for Moon. For centuries, it was widely believed that exposure to moonlight, especially on a full moon night, could alter human behavior.

Aristotle and Pliny the Elder suggested that moonlight influenced fluids in the brain, just as it moved the tides.

In medieval Europe, "moon madness" was taken seriously; people suffering from epilepsy, sleepwalking, or hysteria were often considered "moonstruck."

## Medical History of Moon Madness

By the 18th century, medical texts classified patients as "lunatics" if their symptoms appeared cyclical, supposedly tied to the moon's phases. Asylums in Britain sometimes restrained patients more during full moons.

Even into the 20th century, police and hospitals noticed spikes in "strange behavior" reports during full moons — though modern statistics suggest this was more myth than reality.

## Science vs. Folklore

Modern science dismisses the moon's direct influence on mental health, except perhaps through sleep disruption caused by brightness. Yet folklore lingers:

Emergency workers still report expecting "busier nights" on full moons.

Crime fiction and horror continue to use lunacy as a motif.

Everyday speech keeps phrases like "moonstruck" alive.

The myth may not be medical truth, but it remains psychological truth in culture.

## Symbolism of the Full Moon in Western Imagination

The full moon's imagery in Western lore often tilts toward instability, transformation, and danger:

Werewolves → symbolizing the beast within humanity.

Lunatics → reflecting society's fear of mental illness.

Witches' Sabbaths → often timed with full moons, when magic was believed most powerful.

Romantic Gothicism → poets like Lord Byron and Shelley described the full moon as both beautiful and terrifying.

The moon thus became a dual image: silver goddess of beauty and merciless eye of madness.

## Case Studies in Folklore

1. The Werewolf of Bedburg (Germany, 1589)

Peter Stumpp, a farmer, was accused of being a werewolf who killed dozens. Under torture, he "confessed" to making a pact with the devil and wearing a magical wolf belt. His execution was brutal, spreading werewolf terror across Europe.

2. Lunatics in Victorian England

Doctors recorded patients going into fits "only when the moon was full." Some asylums introduced moonlight-blocking shutters. Later studies showed these patients suffered from epilepsy or bipolar cycles, not moonlight.

3. The Moonlit Forests of France

Folklore from Brittany warned children never to sleep in moonlight, or they would wake up "half-beast." This story reinforced parental authority while embedding lunar fear in culture.

## The Moon in Art and Storytelling

Western painters and storytellers leaned heavily into moonlit horror:

Gothic art showed the moon as a pale face watching over graveyards.

Victorian novels used the full moon to signal supernatural danger.

Modern cinema ties werewolf transformations to dramatic full moon rises.

The moon thus moved from the agricultural calendar to the theatrical stage of Western imagination.

## Why These Myths Persist

Why do werewolves and lunacy still fascinate us?

Psychological Projection – The moon became a mirror of hidden fears and desires.

Natural Mystery – Its cycle, brightness, and effect on tides felt supernatural before science.

Cultural Drama – The idea of man transforming into beast reflects ongoing struggles with morality, violence, and inner darkness.

The full moon offers a perfect stage: dramatic, luminous, and timeless.

## Closing Thoughts

"The full moon still shines as it did thousands of years ago, but we no longer fear it as a curse. Instead, we inherit the stories it inspired—of wolves, witches, and madness. They remind us of a time when night belonged not to science, but to shadows."

Western fairytales turned the Moon into a symbol of duality: beauty and madness, light and terror, guidance and danger. And even today, when the full moon rises, we still feel a twinge of that ancient unease… as if somewhere, in the dark, a wolf might be watching.

**king punished by Zeus**

# 30 - Moon In Indian Classical And Folk Narratives

## The Eternal Indian Moon

*In Indian culture, the Moon (चंद्र (सोम /has always been more than just a celestial body. It is a deity, a calendar keeper, a poet's muse, a lover's companion, and a folk storyteller's canvas. Unlike the Western association of the Moon with fear or lunacy, in India the Moon is associated with beauty, rhythm, time, devotion, and longing.*

From the Vedic hymns of the Rigveda, to the ornate poetry of Kalidasa, to village lullabies sung by grandmothers in rural India, the Moon shines as an eternal symbol of emotion, devotion, and cosmic order.

## I. Moon in the Vedas and Upanishads

### Soma – The Divine Elixir

In the Rigveda, the Moon is personified as Soma, the sacred drink of immortality.

Soma is described as flowing like silver light, nourishing gods and humans alike.

Rigveda 9.113 praises Soma as:

"Soma, the immortal drop, grants us health, wealth, and wisdom."

The waxing and waning of the Moon was seen as the cyclical drinking of Soma by the gods.

### Chandrama as Time-Keeper

In the Yajurveda, the Moon is identified as the mind of the Cosmic Being (Purusha).

The Taittiriya Samhita connects the Moon with the calendar:

Shukla Paksha → growth, vitality.

Krishna Paksha → decline, withdrawal.

Thus, the Moon became a marker of time and rituals, central to Indian life.

## II. Puranic Narratives

### Chandra – The Moon God

In the Puranas, Chandra is described as a handsome deity, wearing a crown of jewels, riding a chariot pulled by ten white horses or an antelope.

Parents: Sage Atri & Anasuya.

Spouse: 27 daughters of Daksha (the 27 Nakshatras).

Famous Myth: Chandra loved Rohini more than the others. This favoritism caused Daksha's curse, leading to waning of the Moon. Later, Shiva partially relieved him, creating the waxing-waning cycle.

This myth explains both astronomy (lunar phases) and morality (importance of balance & fairness).

### Chandra & Shiva – The Crescent Moon

Lord Shiva wears the crescent Moon in his hair (Chandrashekhara).

Symbolism:

The Moon = Time

Shiva = Eternal beyond time

Hence, Shiva's adornment shows victory over time and death.

### Churning of the Ocean (Samudra Manthan)

During the great churning, many treasures emerged – including Chandra.

He was placed on Shiva's head, establishing him as a cosmic ornament.

## III. Moon in Epics – Ramayana & Mahabharata

**Ramayana**

Valmiki compares Rama's face to a full moon countless times.

In Sita's longing, she gazes at the moonlit night, remembering Rama.

The Moon here is a companion of separated lovers, evoking viraha rasa (emotion of longing).

**Mahabharata**

Chandra is the father of Budha (Mercury) through his union with Tara, Brihaspati's wife.

The epic thus connects lunar and planetary mythology, forming the basis of Jyotisha (astrology).

## IV. Moon in Classical Sanskrit Poetry

**Kalidasa**

No poet celebrated the Moon like Kalidasa.

In Meghaduta, the separated lover describes how the Moon intensifies longing.

In Kumarasambhava, Parvati's beauty is compared to the full Moon.

For Kalidasa, the Moon was both romantic and spiritual.

- Jayadeva's Gita Govinda
- Krishna's face is described as "Chandramukha" (moon-faced).
- Radha's night-long wait is painted with moonlit imagery.

**Sangam Literature (Tamil)**

The Moon is tied to akam poetry (love/inner life).

Lovers separated by distance often confide in the Moon as their messenger.

## V. Moon in Indian Folk Traditions

**Rural Songs & Lullabies**

In Maharashtra, lullabies call the Moon "Chanda mama" (Uncle Moon).

In Bengal, children's rhymes invite "Chand mama ashe pore ronger ghore".

Across villages, the Moon is a friendly relative who brings comfort to children.

**Harvest & Festivals**

Kojagiri Purnima – Families worship the full Moon goddess, seeking prosperity.

Karva Chauth – Women break their fast after sighting the Moon through a sieve.

Chhath Puja (Bihar) – Worship of Sun also includes offering arghya to the setting and rising Moon.

## VI. Tribal Narratives and Folklore

**Gond Tribe**

Believe the Moon is a bride visiting her husband (the Sun).

Their paintings often depict Moon phases as parts of this union.

**Santhal Tribe**

Narratives describe the Moon as a protector of night travelers.

Some rituals invoke Moon to ward off evil spirits.

**Nagaland & North-East India**

Folklore links fertility cycles of women with the Moon.

Agricultural sowing often aligned with new/full moon nights.

## VII. Moon in Festivals & Rituals

Holi Purnima – marks transition of seasons.

Guru Purnima – dedicated to teachers, aligned with Moon's fullness (symbol of enlightenment).

Sharad Purnima – considered night of divine nectar (Amrit Varsha).

Buddha Purnima – birth, enlightenment, and nirvana of Buddha all under full Moons.

## VIII . Psychological & Emotional Symbolism

Love & Longing – Moonlit metaphors dominate romantic literature.

Peace & Comfort – lullabies invoke Moon for calmness.

Time & Discipline – Moon cycles regulate festivals, rituals, and fasts.

Mysticism – Yogis meditate under full Moon nights, believing it enhances energy.

## Conclusion – The Eternal Moon of India

The Indian Moon is not a cold, distant satellite, but a living presence – a god, a relative, a poet's metaphor, and a villager's comfort.

From Vedic Soma to Daksha's cursed Chandra, from Kalidasa's poetry to tribal songs, the Moon is woven deeply into India's spiritual, literary, and folk heritage.

Unlike the West's fearful werewolves, the Indian Moon inspires devotion, beauty, and rhythm of life.

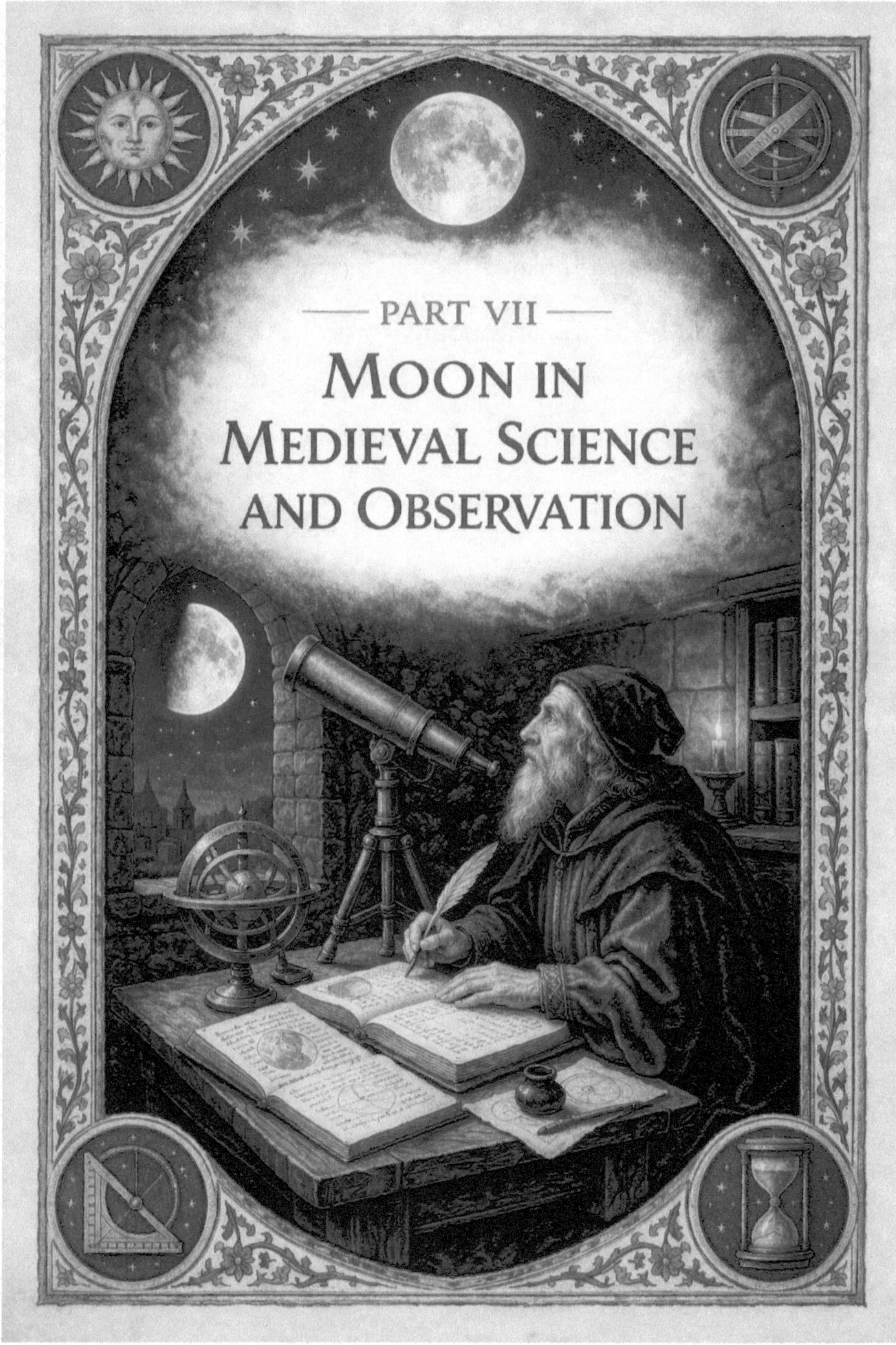
PART VII
MOON IN
MEDIEVAL SCIENCE
AND OBSERVATION

# 31- Aryabhata, Varahamihira, And Indian Lunar Models

## India's Scientific Legacy and the Moon

Why the Moon fascinated Indian astronomers and mathematicians.

The role of Jyotisha Shastra (astronomy + astrology) in Vedic & classical India.

Transition from ritual calendars to mathematical astronomy.

## I. Aryabhata (476–550 CE) The Pioneer of Indian Astronomy

### Life & Historical Context

Born in Kusumapura (modern Patna).

His magnum opus: Aryabhatiya (composed in 499 CE).

### Aryabhata's Lunar Theories

Geocentric Model with Rotating Earth

Proposed that Earth rotates on its axis daily (revolution concept missing, but major leap).

Explained lunar phases through sunlight reflection – rejecting mythological shadows.

### Calculation of Synodic & Sidereal Months

Sidereal month = ~27.321 days (very close to modern 27.321661).

Synodic month = ~29.5306 days (modern = 29.53059).

### Eclipses Explained Scientifically

Aryabhata rejected "Rahu–Ketu swallowing Moon" mythology.

Correct explanation: Earth's shadow causes lunar eclipses.

**Mathematical Precision**

Aryabhata computed pi ($\pi \approx 3.1416$).

Calculated Moon's mean distance using trigonometry.

Aryabhata's Influence on Arab & Europe

Translated into Arabic as Al-Arjabhar.

His lunar models shaped Islamic calendars & European medieval astronomy.

## II. Varahamihira (505–587 CE) – The Polymath

**Biography**

Born in Ujjain – India's astronomical hub.

Masterpiece: Brihat Samhita (encyclopedic work).

**Lunar Contributions**

Nakshatra System Expanded

Linked lunar motion to 27 constellations.

System used in Hindu astrology till today.

**Calendrical Reforms**

Corrected tithi, eclipses, and intercalary months.

Built practical almanacs for kings & farmers.

**Cultural Integration**

Moon linked with rainfall prediction.

Festivals, agriculture, and astrology blended with lunar observations.

**Varahamihira's Legacy**

Combined mathematics, astronomy, meteorology, astrology, and omens.

Gave Moon a civilizational role, not just a scientific one.

## III. Other Indian Thinkers and Lunar Models

**Brahmagupta (598–668 CE)**

Gave precise equations for lunar eclipses.

Corrected Aryabhata's models with better mean values.

**Bhaskaracharya (1114–1185 CE)**

Siddhanta Shiromani – lunar equations, planetary positions.

Discussed Moon's parallax and "Chandramandal".

**Kerala School (14th–16th century)**

Madhava of Sangamagrama developed infinite series for lunar calculations.

Anticipated modern calculus.

## IV. Indian Lunar Models – Science vs Myth

Indian tradition balanced astronomy (Ganita Jyotisha) with astrology (Hora Shastra).

While Aryabhata explained eclipses scientifically, common people still believed Rahu–Ketu myths.

This dual tradition survived – both scientific siddhantas and mythological puranas co-existed.

## V. Global Influence of Indian Lunar Science

Arabic & Islamic Astronomy

Aryabhata and Brahmagupta's lunar tables translated into Arabic.

Helped shape Islamic Hijri lunar calendar corrections.

Chinese Influence

Indian Siddhantas introduced to Tang dynasty astronomers.

European Medieval Astronomy

Islamic translations carried Indian lunar models to Europe.

Influenced Kepler's tables indirectly.

## VI. Practical Applications in Indian Life

Agriculture – Planting & harvesting based on lunar phases.

Festivals – Diwali (Amavasya), Holi (Purnima).

Temple rituals – Panchangam = Moon-based.

Medicine (Ayurveda) – Some therapies timed with lunar cycles.

## VII. Conclusion – The Indian Moon Legacy

Aryabhata introduced mathematical astronomy.

Varahamihira integrated cosmos with culture.

Indian lunar models became global heritage, influencing Islam, China, and Europe.

Even today, the Hindu Panchang, Vedic astrology, and scientific astronomy carry their legacy forward.

Aryabhata
Varahamihira

# Chapter 32: Galileo Galilei's Moon Sketches

## A Turning Point in Astronomy

Before Galileo, the Moon was seen as a perfect celestial sphere, smooth and flawless, reflecting divine harmony. This belief came from Aristotle and Ptolemy's models, and was strongly supported by the Church.

But in 1609–1610, Galileo Galilei, armed with a newly improved telescope, challenged this age-old vision. His sketches of the Moon, published in Sidereus Nuncius (The Starry Messenger, 1610), forever changed how humanity saw the cosmos.

## I. The Telescope and Galileo's Curiosity

Galileo did not invent the telescope, but he enhanced its design, making lenses that could magnify objects nearly 20 times.

Turning it to the Moon in late 1609, he was astonished.

Instead of a flawless orb, he saw shadows, irregular edges, mountains, and craters.

For the first time, a human being realized:

*The Moon was a landscape, not a perfect crystal sphere.

## II. The Sketches – Drawing the Face of the Moon

Galileo made more than just observations; he sketched what he saw.

His drawings showed dark and light patches, which he interpreted as mountains, valleys, and plains.

He carefully noted how shadows changed as the Moon moved through its phases.

He even estimated the heights of lunar mountains (up to 4 miles high), using simple geometry of light and shadows.

These sketches appeared in Sidereus Nuncius (1610), making it one of the first printed books illustrated with telescopic observations.

Early sketches show a half-Moon with dark shading near the terminator (the line dividing light and darkness).

He compared the Moon's mountains to the Apennine mountains of Italy.

### III. Breaking the Myth of Perfection

The idea of the Moon as "corruptible and imperfect" was revolutionary.

Aristotle's cosmology said the heavens were made of aether—perfect and unchanging.

Galileo's sketches directly contradicted this.

The Moon was a world, much like Earth.

This had profound implications:

If the Moon was Earth-like, could other celestial bodies be similar?

Was Earth not unique, but just one of many planets?

### IV. Scientific Method in Action

Galileo used mathematics and observation together.

By measuring the angle of sunlight on lunar mountains, he calculated their approximate heights.

He proved that science was not just philosophy, but could use experiments and geometry to reveal truths.

His work represented the shift from belief to empirical evidence.

### V. Publication – Sidereus Nuncius (The Starry Messenger)

Published in March 1610, it contained:

- Sketches of the Moon

- Observations of Jupiter's four largest moons (Io, Europa, Ganymede, Callisto)
- Studies of countless stars invisible to the naked eye

This book shook Europe. Scholars, philosophers, and clergy were stunned.

### ×VI. Resistance and Controversy

The Church and Aristotelian scholars opposed Galileo's findings.

They argued that the telescope was deceptive, or that imperfections on the Moon were illusions.

Some refused to look through his telescope at all, fearing it contradicted religious doctrine.

Yet, Galileo's evidence was undeniable.

### VII. Impact on Future Science

The Moon became a physical object – opening the way for lunar science.

Planetary geology began with Galileo's sketches.

His methods inspired Kepler, Newton, and later astronomers.

Today, NASA scientists still consider Galileo's drawings the first step toward lunar exploration.

### VIII .Philosophical Legacy

Galileo's Moon sketches were not just drawings. They symbolized:

The death of Aristotelian perfection.

The birth of modern observational astronomy.

Humanity's first real step toward seeing the universe as physical and measurable, not mystical and untouchable.

### Conclusion

When Galileo pointed his telescope at the Moon, he changed not only astronomy but also human imagination.

The Moon was no longer a perfect heavenly lamp, but a cratered, mountainous world, perhaps even like Earth.

His sketches are among the most important scientific illustrations ever made, marking the dawn of the scientific revolution.

*"The surface of the Moon is not smooth, uniform, and precisely spherical, as a great number of philosophers believe it… but, on the contrary, it is uneven, rough, and crowded with depressions and bulges, just like the surface of the Earth."*

**– Galileo Galilei, Sidereus Nuncius (1610)**

Hæc eadem macula ante ſecundam quadraturam nigrioribus quibuſdam terminis circumuallata conſpicitur; qui tanquam altiſſima montium iuga ex parte Soli auerſa obſcuriores apparent, quà verò Solem reſpiciunt lucidiores extant; cuius oppoſitum in cauitatibus accidit, quarum pars Soli auerſa ſplendens apparet, obſcura verò, ac vmbroſa, quæ ex parte Solis ſita eſt. Imminuta deinde luminoſa ſuperficie, cum primum tota fermè dicta macula tenebris eſt obducta, clariora mõtium dorſa eminenter tenebras ſcandunt. Hanc duplicem apparentiam ſequentes figuræ commonſtrant.

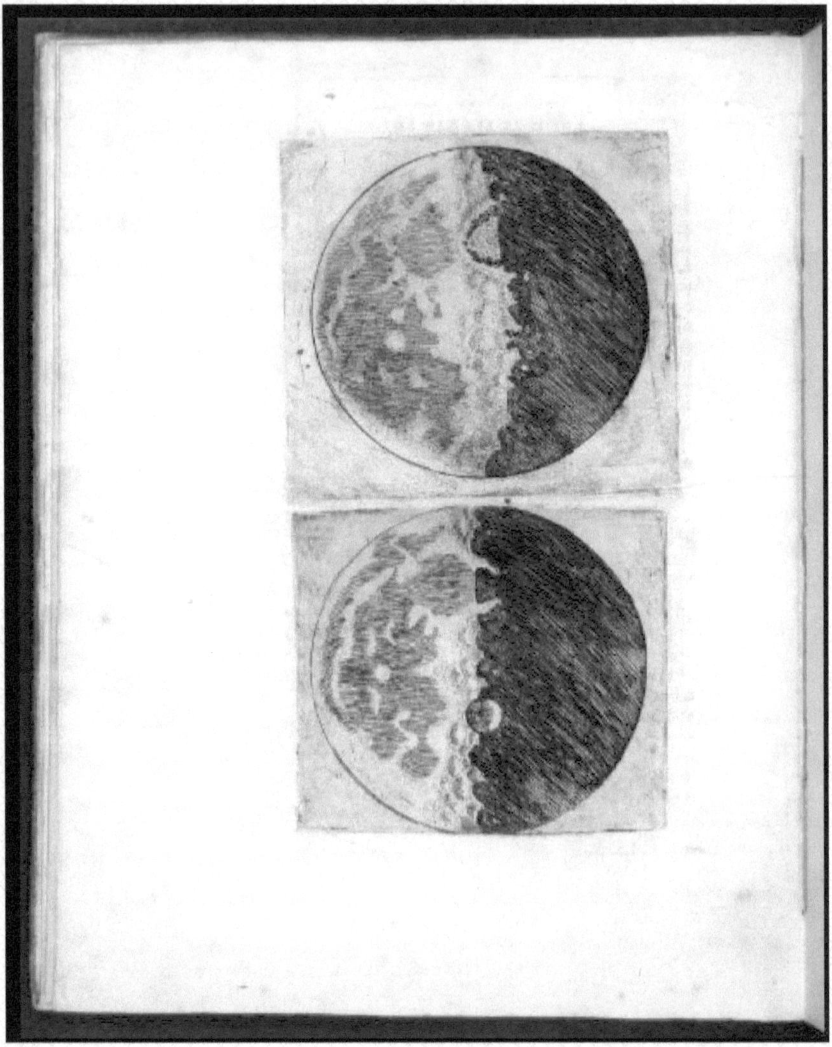

# 33 - Islamic Astronomers And Lunar Models

***When Europe was still bound by Aristotelian cosmology, the Islamic world became a hub of scientific curiosity. From the 8th to the 15th century, scholars in Baghdad, Damascus, Cairo, Cordoba, and Samarkand advanced astronomy by combining Greek, Indian, and Persian knowledge. The Moon, being central to Islamic ritual life (Ramadan, Eid, Hijri calendar), was studied with great care. This chapter explores how Islamic astronomers refined lunar models, built observatories, and influenced later European science.***

## I. Astronomy in the Islamic Golden Age

The Abbasid Caliphs (8th–13th century) established the Bayt al-Hikma (House of Wisdom) in Baghdad.

Ancient Greek texts (Ptolemy's Almagest), Indian Siddhantas, and Persian Zij works were translated into Arabic.

The institution was destroyed during the Mongol siege of Baghdad in 1258, during which its books were famously thrown into the Tigris River.

Astronomy was not purely theoretical – it had religious necessity:

- Determining prayer times (salat).
- Establishing the direction of Mecca (qibla).
- Regulating the lunar Islamic calendar.

## II. Al-Khwarizmi and the First Islamic Lunar Tables

Muhammad ibn Musa al-Khwarizmi (c. 780–850) created one of the earliest Islamic astronomical tables.

His Zij al-Sindhind incorporated Indian and Persian data.

Lunar cycles (synodic, sidereal, anomalistic) were calculated to improve prediction of new Moons.

His works later influenced Latin scholars, giving Europe the word **"algorithm".**

### III. Al-Battani – The Refiner of Lunar Orbits

Al-Battani (858–929), also known as Albatenius in Latin, corrected Ptolemy's lunar parameters.

He refined the length of the solar year and gave more accurate values for the Moon's apogee and inclination.

His lunar tables were so precise that Copernicus quoted him 700 years later.

He described how the Moon's apparent size changes due to its elliptical orbit – a key step toward understanding lunar anomalies.

### IV. Ibn Yunus and Large-Scale Observations

Ibn Yunus (950–1009) in Cairo used massive astronomical instruments (may be a giant mural quadrant).

He compiled over 10,000 observations of the Sun and Moon.

His lunar models corrected Ptolemaic values with better precision, laying groundwork for later trigonometric astronomy.

His accuracy in predicting eclipses was extraordinary for his era.

### V. Lunar Astronomy and Islamic Rituals

Islamic law required eyewitness sighting of the crescent (hilal) to mark Ramadan and Eid.

Scholars debated whether calculation or observation was superior.

While jurists preferred observation, astronomers developed advanced lunar ephemerides to predict new Moons.

This tension between religion and science shaped the unique path of Islamic astronomy.

## VI. Al-Biruni – The Philosopher-Scientist

Al-Biruni (973–1048) combined observational data with mathematical theory.

He described the Moon's phases, eclipses, and parallax in his works.

He proposed the Earth rotated on its axis and attempted to measure Earth's circumference.

In lunar studies, he discussed how shadows on the Moon suggested mountains and valleys – anticipating Galileo centuries later.

## VII. Nasir al-Din al-Tusi and the Tusi Couple

At the Maragha Observatory (1259, Persia), Nasir al-Din al-Tusi (1201–1274) developed the famous Tusi couple (a mathematical device).

This mechanism allowed linear oscillations to be generated from circular motions.

It solved some of the inconsistencies in Ptolemy's lunar model.

Later, Copernicus adopted the Tusi couple in his De Revolutionibus.

## VIII. Ulugh Beg and the Samarkand Observatory

Ulugh Beg (1394–1449), Timurid prince and astronomer, built one of the largest observatories of the medieval world in Samarkand.

His Zij-i Sultani included detailed lunar and planetary data.

The Moon's motions were cataloged with remarkable accuracy, using giant sextants and quadrants.

His work represented the pinnacle of Islamic observatory-based astronomy.

## IX. Influence on Europe

Latin translations of Islamic works entered Europe via Spain and Sicily.

Astronomers like Gerard of Cremona translated Almagest commentaries with Islamic refinements.

Copernicus, Kepler, and even Galileo drew upon Islamic sources.

Lunar models that started in Baghdad ended up shaping the European Scientific Revolution.

### Conclusion

Islamic astronomers bridged ancient wisdom and modern science. Their lunar models corrected errors in Ptolemaic astronomy, laid mathematical foundations, and preserved knowledge that Europe later inherited.

The Moon was both a sacred calendar keeper and a scientific challenge. By measuring its motions with rigor, Islamic astronomers advanced not only religious practice but also global science. Without their contributions, the telescope era of Galileo would have stood on much shakier ground.

A Soviet postage stamp issued 6 September 1983, commemorating al-Khwārizmī's (approximate) 1200th birthday

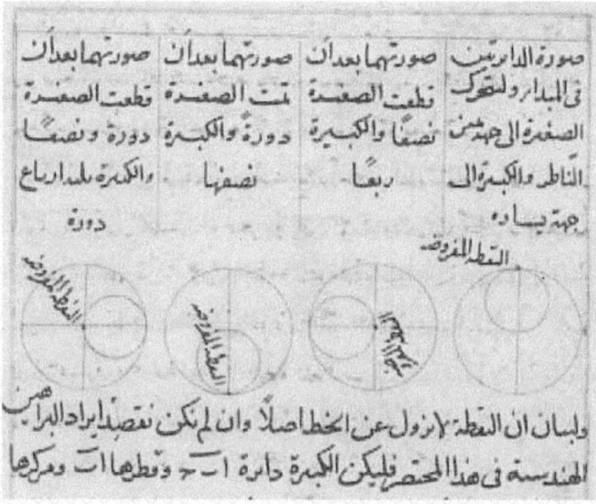

Tusi's diagram of the Tusi couple, 13th century

“A model of Ulug Beg Observatory in the Uzbek National Museum” by **Victoria**, licensed under **Creative Commons Attribution-ShareAlike 4.0 International (CC BY-SA 4.0)** / Wikimedia Commons.

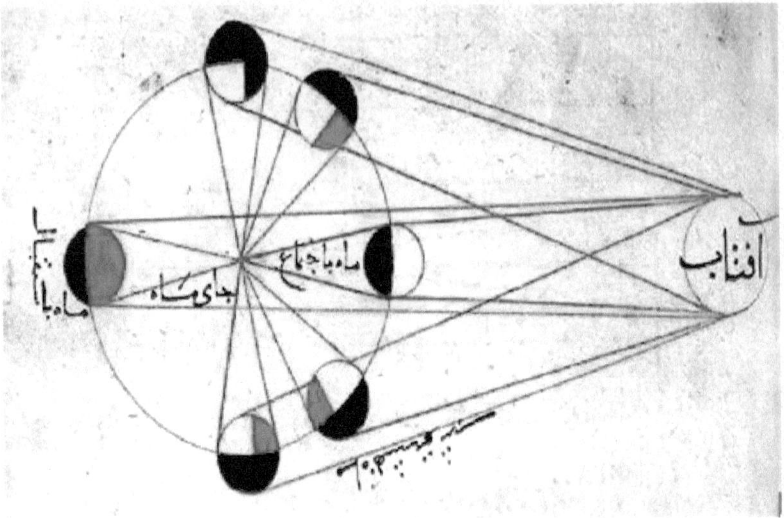


*Astronomical diagram illustrating lunar phenomena* from a manuscript of **Abu al-Rayhan al-Biruni** (973–1048). Medieval Islamic manuscript. **Public domain** (faithful reproduction), via Wikimedia Commons.

# 34 - European Lunar Maps And Navigation

## From Mystical Moon to Measured Globe

*By the 16th and 17th centuries, Europe was entering the Age of Exploration. Mariners sought new routes across the seas, while astronomers sought precision in the skies. The Moon—long a symbol of myth and mystery—was now being re-imagined as a practical tool for science, cartography, and navigation. Lunar maps became crucial not only for astronomy but also for sailors, kings, and scientists who demanded accuracy in navigation and timekeeping.*

## I. Early European Lunar Cartography

Before Galileo (pre-1600), lunar maps were crude drawings in manuscripts. The Moon was still seen as symbolic, representing cycles of time, femininity, or divine order.

After Galileo (1610), telescopic observations revealed mountains, craters, and plains. Artists and astronomers began sketching these features systematically.

Galileo's Sidereus Nuncius contained the first realistic lunar drawings, but they were more artistic than cartographic.

## II. The First True Lunar Maps

i) Michael van Langren (Langrenus Map, 1645)

Flemish engineer and cartographer.

Published the first named lunar map, assigning Christian names (often of Catholic saints and kings) to craters and mountains.

Example: Mare Crisium was named Mare Philippicum after King Philip IV of Spain.

His map was also political propaganda, embedding Catholic influence in the sky.

ii) Johannes Hevelius (Selenographia, 1647)

Polish astronomer, often called the "Father of Lunar Cartography."

Created detailed lunar maps using his own telescope.

Introduced Latinized geographic-style names: Montes Apenninus, Mare Tranquillitatis.

His maps treated the Moon like a new continent to be explored.

Hevelius also invented the word selenography (science of mapping the Moon).

iii)Giovanni Battista Riccioli (Almagestum Novum, 1651)

Italian Jesuit astronomer.

Developed the naming system still in use today:

- Craters named after philosophers, scientists, and astronomers (e.g., Plato, Archimedes, and Copernicus).
- Dark plains called Mare (seas).

His system was more neutral and gained international acceptance.

## III. The Moon as a Tool for Navigation

In the 17th and 18th centuries, one of the greatest challenges for sailors was determining longitude at sea.

Latitude was easy (measured by the Sun's height), but longitude required knowing the exact time difference between two locations.

The Moon became the solution:

**The Lunar Distance Method**

Proposed in the 16th century, refined in the 18th.

Sailors measured the angular distance between the Moon and a known star or planet.

Using tables (like the Nautical Almanac, first published in 1767), they could calculate Greenwich time and thus their longitude.

Accuracy depended heavily on detailed lunar maps and predictions of lunar motion.

## IV. Key European Contributions

### Tobias Mayer (1750s)

German astronomer who created highly accurate lunar tables.

His data enabled sailors to calculate longitude within half a degree.

Awarded by the British Parliament as part of the Longitude Prize.

### Royal Greenwich Observatory

Founded in 1675 by King Charles II of England.

Its primary mission: improve lunar tables for navigation.

Astronomers like John Flamsteed and Nevil Maskelyne refined lunar predictions.

### The Nautical Almanac (1767 onwards)

Published annually in Britain.

Contained lunar positions and distances, allowing sailors worldwide to use the lunar method.

For over a century, this was the standard navigation tool until marine chronometers became affordable.

## V. The Moon in European Art and Exploration

The Moon was often illustrated in early atlases and celestial globes.

European explorers considered the Moon both a guide for navigation and a symbol of conquest—just as they mapped new lands, they sought to map the skies.

By the 18th century, engravings of lunar maps circulated widely, inspiring both scientific communities and the general public.

## VI. From Cartography to Lunar Science

These early European lunar maps and navigation methods laid the groundwork for:

Planetary geology (treating the Moon as a terrain with mountains, plains, and valleys).

Space navigation centuries later (Apollo astronauts in the 20th century relied on improved versions of selenography).

The tradition begun by Langrenus, Hevelius, and Riccioli marked the transition from myth to measurement, from art to science.

## Conclusion

The Moon, once only a mystical lamp of the night, became a measured globe in European science.

Through detailed lunar maps and the ingenious use of the Moon for navigation, European astronomers transformed both astronomy and seafaring. These contributions ensured safe ocean voyages, advanced scientific accuracy, and created a legacy that still echoes in modern lunar missions.

The story of European lunar maps is not merely about the Moon—it is about humanity's desire to explore, measure, and master both Earth and sky.

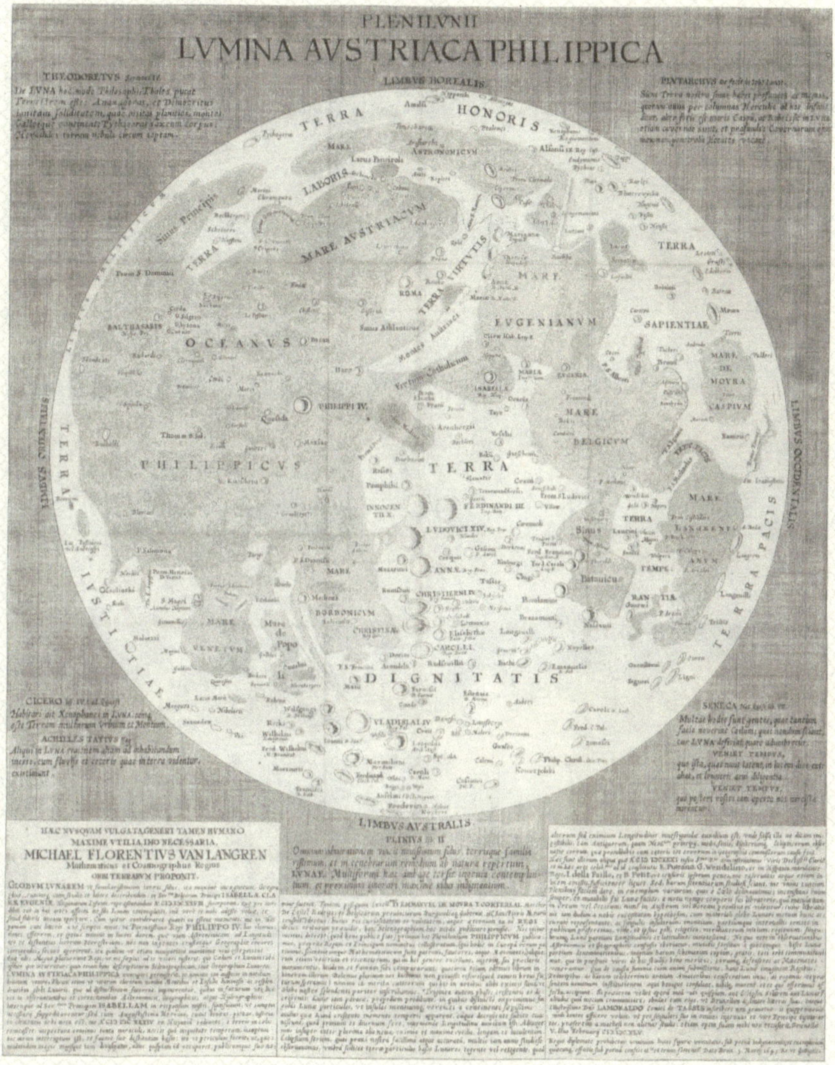


Map of the Moon, published by **Michael van Langren** in 1645 in his book *Plenilunii lumina Austriaca Philippica*

Johannes Hevelius
Portrait by Daniel Schultz

# 35 - Astrology Vs. Astronomy Debate In The Middle Ages

## The shared origins of astrology and astronomy

In the modern era, astrology and astronomy are considered two very different disciplines—one is seen as a pseudoscience, the other as a legitimate science. Yet in the Middle Ages, these fields were deeply intertwined. Both emerged from the same roots in Babylonian, Greek, and Indian traditions, where observing the stars served practical and mystical purposes alike.

For centuries, the sky was studied to mark time, predict weather, guide navigation, and divine human destiny. The same star charts were used by priests for rituals and by astronomers for measuring celestial motion. The split between astrology and astronomy began slowly, influenced by philosophy, religion, and the rise of empirical science. The Middle Ages (roughly 500–1500 CE) was the key battleground of this intellectual debate.

## I. Historical Background: Before the Middle Ages

**Babylonian Astronomy-Astrology (2000 BCE onwards):**

- Priests of Mesopotamia recorded planetary positions on clay tablets.
- Eclipses and planetary conjunctions were linked to omens for kings and empires.

**Greek Contributions:**

- Claudius Ptolemy's Tetrabiblos (astrology) and Almagest (astronomy) became foundational texts.
- For Ptolemy, astrology was not superstition but a natural extension of astronomy.

**Indian and Islamic Worlds:**

- Indian Jyotisha (astrology) and Nakshatra astronomy influenced Arab scholars.
- Islamic Golden Age thinkers translated Greek works and developed astronomical tables (zij).

By the time these traditions reached medieval Europe, astrology and astronomy were still considered two sides of the same coin.

## II. Astrology in the Medieval World

Astrology was widely practiced in the Middle Ages, not as fringe mysticism but as a respected intellectual pursuit.

**Universities and Astrology:**

In the 13th and 14th centuries, astrology was part of the university curriculum in Paris, Bologna, and Oxford.

Medicine and astrology were linked: physicians studied planetary positions to diagnose illness and decide treatments.

**Royal Courts:**

Kings and nobles employed astrologers to choose auspicious dates for battles, coronations, and marriages.

Example: Emperor Frederick II (1194–1250) encouraged astrological studies in Sicily.

**Religious Ambivalence:**

The Church allowed astronomical astrology (e.g., using stars to tell seasons) but condemned deterministic astrology (belief that planets control free will).

Augustine of Hippo strongly opposed astrology, arguing that it undermined Christian doctrine of free will.

## III. Astronomy in the Middle Ages

Astronomy, though less glamorous than astrology, developed steadily through observatories, instruments, and translations.

**Islamic Golden Age:**

Scholars like Al-Battani (Albategnius), Al-Sufi, and Al-Tusi made accurate star catalogues and planetary models.

They refined Ptolemy's work and passed knowledge to Europe via Spain and Sicily.

**European Astronomy:**

The 12th-century translation movement in Toledo brought Arabic and Greek texts into Latin.

Monks studied astronomy to calculate Easter and maintain liturgical calendars.

**Instruments:**

Astrolabes, quadrants, and armillary spheres were used both by astronomers and astrologers.

Navigation improved with these tools, linking astronomy directly to exploration.

## IV. The Debate: Astrology vs. Astronomy

The tension grew because both fields shared methods but had different goals.

**Astronomy's Goal:**

To measure and predict celestial motions with accuracy.

Grounded in mathematics and geometry.

**Astrology's Goal:**

To interpret those motions as signs influencing human fate.

Grounded in symbolic and metaphysical reasoning.

**Core Points of Debate:**

Determinism vs. Free Will

Astrologers argued stars shaped destiny.

Theologians and some philosophers said this denied God's providence and human choice.

Empirical vs. Symbolic

Astronomy tested predictions with observation.

Astrology relied on symbolic associations (e.g., Mars = war).

Practical Usefulness

Astronomers were needed for calendars, navigation, agriculture.

Astrologers claimed relevance to medicine, politics, and personal life.

## V. Case Studies from the Middle Ages

1. Albertus Magnus (1200–1280)

Dominican friar and scholar.

Accepted astrology as a natural science but rejected fatalistic astrology.

2. Roger Bacon (1219–1292)

English philosopher who valued astronomy for navigation.

Criticized astrology when it lacked rational proof.

3. Thomas Aquinas (1225–1274)

Argued that stars might influence the human body but not the soul.

Allowed limited astrological influence, subordinated to God's will.

4. Dante Alighieri (1265–1321)

In Divine Comedy, he used astronomical imagery but warned against trusting astrology.

## VI. Popular Belief vs. Scholarly Debate

Among common people, astrology remained highly popular.

Almanacs predicting weather, harvests, and personal fortunes were widely sold.

Astrology was even used to explain plagues and comets.

Astronomers were often forced to produce astrological charts to earn patronage.

## VII. The Church and Astrology

The Catholic Church took a nuanced position:

Condemned deterministic astrology in 1277 Condemnations at Paris.

But Popes often kept court astrologers! (e.g., Pope Sixtus IV consulted astrologers).

This hypocrisy shows the ambiguity of medieval society toward astrology.

## VIII. Turning Point – Renaissance & Scientific Revolution

By the late Middle Ages, cracks appeared:

Copernicus (1473–1543):

His heliocentric theory (1543) shook astrology's framework, which was Earth-centered.

Tycho Brahe (1546–1601):

Practiced astrology but emphasized precise measurements.

Johannes Kepler (1571–1630):

Cast horoscopes for income but rejected symbolic astrology.

Used astronomy to discover planetary laws.

Thus, the Renaissance transformed astrology from "serious science" to "superstition," while astronomy became the cornerstone of modern science.

## IX. Legacy of the Debate

The Middle Ages debate shaped intellectual culture:

Preserved ancient wisdom through astrology.

Strengthened scientific method through astronomy.

Even today, horoscopes remain popular, showing astrology's cultural resilience.

Meanwhile, astronomy went on to lead space exploration, satellites, and cosmology.

## Conclusion

The Astrology vs. Astronomy debate in the Middle Ages was not a simple clash of reason vs. superstition. It was a dialogue between two intertwined traditions, each reflecting humanity's desire to find meaning in the heavens.

Astrology gave symbolism and hope; astronomy gave precision and knowledge. Together, they built the intellectual foundation of the Renaissance and Scientific Revolution.

In many ways, medieval astrology kept alive the interest in the stars until astronomy was strong enough to stand on its own. The split was painful, but it gave birth to modern science—while astrology survived as a cultural and psychological system.

*"The stars incline, they do not compel." – Medieval proverb*

Dante reading the *Divine Comedy* at the court of Guido Novello; painting by Andrea Pierini, 1850 (Palazzo Pitti, Florence)

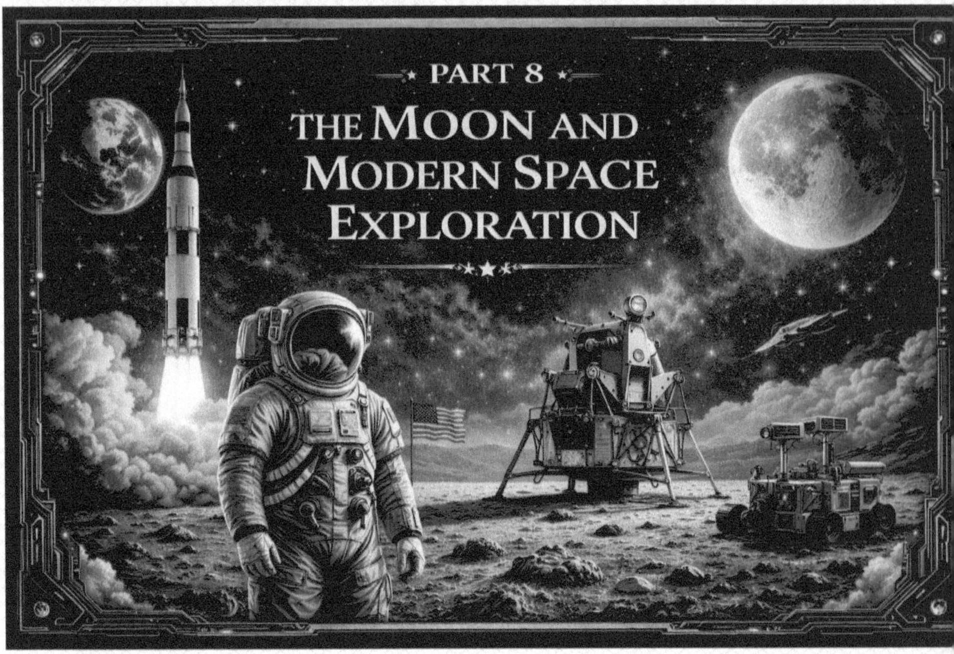
PART 8
THE MOON AND
MODERN SPACE
EXPLORATION

# 36 - Space Race – Apollo Missions And Neil Armstrong

## Cold War Rivalry and the Birth of the Space Race

The Space Race was not just about rockets or science—it was a battle of prestige, ideology, and global leadership. After World War II, the world was divided between two superpowers: the United States (promoting democracy and capitalism) and the Soviet Union (advocating communism and state control). Each side wanted to prove its system was superior.

***Space became the ultimate arena of this rivalry. Achievements beyond Earth would inspire not only national pride but also international influence. Whoever conquered the skies—and later, the Moon—would win the imagination of the entire world.***

The launch of Sputnik 1 by the Soviet Union in 1957 marked the beginning of this cosmic rivalry. The U.S. was shocked and rushed to respond. This contest would culminate in the Apollo missions, with Neil Armstrong's first step on the Moon (1969) becoming one of the greatest moments in human history.

## I. Early Soviet Triumphs – Lighting the Fire of Competition

- Sputnik 1 (1957): First artificial satellite launched by the USSR.
- Laika (1957): First living being (a dog) sent into orbit.
- Yuri Gagarin (1961): First human in space, aboard Vostok 1.

These successes shook the United States, creating fears that Soviet technology could dominate both space and military power.

## II. America Responds – NASA and the Apollo Program

To counter Soviet dominance, the United States established the National Aeronautics and Space Administration (NASA) in 1958.

- President John F. Kennedy's famous 1961 speech committed America to the ambitious goal: "We choose to go to the Moon in this decade… not because it is easy, but because it is hard."

Thus began the **Apollo Program**—a series of missions designed to master space travel and ultimately land humans on the Moon.

## III. The Apollo Missions – Step by Step to the Moon

### Apollo 1 (1967) – Tragedy

- A cabin fire during a ground test killed astronauts Gus Grissom, Ed White, and Roger B. Chaffee.
- The disaster led to major redesigns in spacecraft safety.

### Apollo 7–10 – Testing the Waters

- Apollo 7 (1968): First successful crewed Apollo mission in Earth orbit.
- Apollo 8 (1968): First humans to orbit the Moon—an emotional triumph. The astronauts read from Genesis on Christmas Eve, watched by millions on Earth.
- Apollo 10 (1969): Dress rehearsal for the Moon landing, testing lunar module operations without landing.

### Apollo 11 (1969) – The Historic Landing

- Date: July 16–24, 1969.
- Crew: Neil Armstrong (Commander), Buzz Aldrin (Lunar Module Pilot), Michael Collins (Command Module Pilot).
- On July 20, 1969, Armstrong and Aldrin landed the lunar module **Eagle** on the Moon's Sea of Tranquility.

Armstrong's first step came with the immortal words:

*"That's one small step for man, one giant leap for mankind."*

They spent about 21 hours on the lunar surface, collecting rocks, setting up experiments, and planting the American flag.

### Later Apollo Missions

- Apollo 12 (1969): Precision landing near Surveyor spacecraft.
- Apollo 13 (1970): "Houston, we've had a problem." A near-disaster after an oxygen tank explosion; crew survived thanks to ingenuity.
- Apollo 15–17 (1971–72): Scientific missions, included lunar rovers for surface exploration.
- Apollo 17 (1972): Last crewed mission to the Moon, with geologist Harrison Schmitt.

## IV. Neil Armstrong – The Man Who Took the First Step

- Born in 1930 in Ohio, USA, Armstrong developed a passion for flying early.
- He became a naval aviator, then a test pilot, flying cutting-edge aircraft.
- Selected as an astronaut in 1962, he was known for his calmness under pressure.
- During Apollo 11, Armstrong manually piloted the lunar module to avoid a boulder field, landing with only seconds of fuel left.

He became a global hero, yet remained humble, often saying he was just doing his job.

## V. Impact of the Apollo Program

### Scientific Discoveries

- Brought back 382 kg of lunar rocks.
- Helped scientists understand the Moon's age, volcanic activity, and relation to Earth.

### Technological Advancements

- Innovations in computing, materials science, and telecommunications.
- Integrated circuits developed for Apollo became precursors to the microchip revolution.

### Cultural Impact

- Inspired a generation of scientists and dreamers.
- The "Earthrise" photo from Apollo 8 highlighted Earth's fragility, boosting the environmental movement.

### ×VI. The Cold War Victory

Landing on the Moon was more than science—it was politics.

- The U.S. demonstrated technological superiority over the Soviet Union.
- While the USSR achieved many firsts, the Moon landing became the symbolic victory of the U.S. in the Space Race.

### VII. Legacy of Apollo and Neil Armstrong

- Apollo proved that humanity could dream big and succeed.
- It remains one of the most complex, daring, and inspiring projects ever undertaken.
- Neil Armstrong's words and footprints are etched forever in human history.

Even today, whenever new missions to the Moon are planned—by NASA's Artemis program or private companies—the legacy of Apollo and Armstrong looms large.

### Conclusion

The Space Race was mor e than a Cold War rivalry. It was the story of human courage, curiosity, and determination to go beyond the limits of Earth. The Apollo missions, crowned by Neil Armstrong's first step, remain one of humanity's greatest achievements—a reminder that when we set our sights high, even the Moon is within our reach.

"The important achievement of Apollo was demonstrating that humanity is not forever chained to this planet." - Neil Armstrong

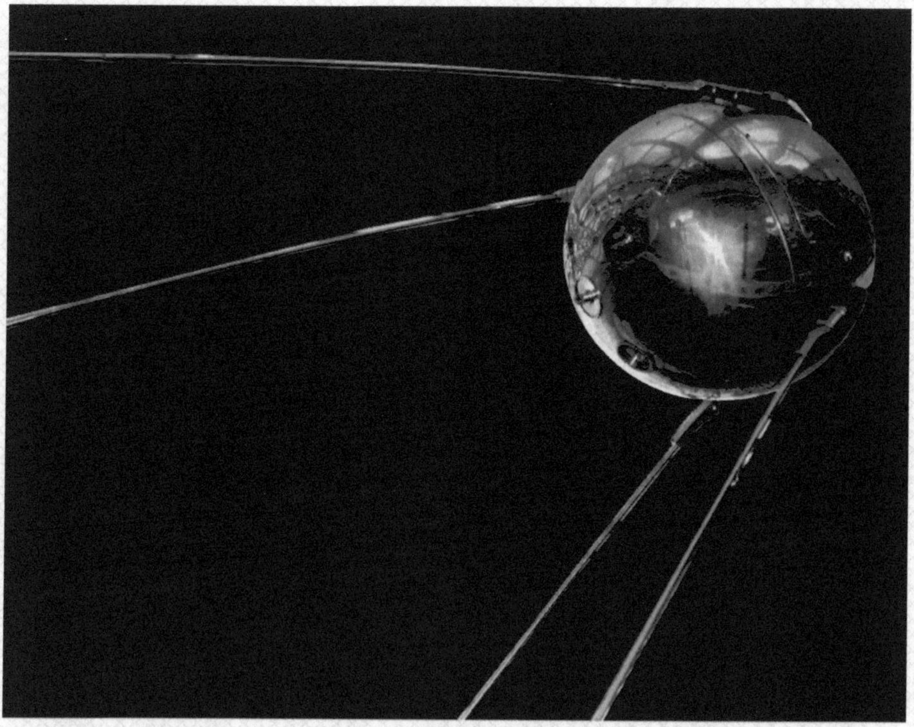

A replica of Sputnik 1, the first artificial satellite in the world to be put into outer space: the replica is stored in the National Air and Space Museum.

Portrait of Astronaut Neil A. Armstrong, commander of the Apollo 11 Lunar Landing mission in his space suit, with his helmet on the table in front of him. Behind him is a large photograph of the lunar surface.

NASA Photo ID: S69-31741 - Program: Apollo XI

# 37 - Lunar Probes – Luna, Chandrayaan, Chang'e

## From Humans to Robots

After the Apollo missions, direct human travel to the Moon became rare due to high costs and risks. But the Moon remained a key scientific and strategic target.

Thus began the era of robotic lunar probes, launched by different nations to study the Moon's surface, composition, and potential for future colonization.

The three most important programs in this journey are:

- Soviet Luna Program (1959–1976)
- India's Chandrayaan Missions (2008–present)
- China's Chang'e Program (2007–present)

Each has contributed uniquely to humanity's knowledge of the Moon.

## I. Soviet Union's Luna Program (1959–1976)

### 1. The First Attempts

Luna 1 (1959): First spacecraft to escape Earth's gravity and fly past the Moon.

Luna 2 (1959): First human-made object to crash into the Moon.

Luna 3 (1959): Sent back the first photographs of the far side of the Moon, a region never seen before.

### 2. Major Achievements

Luna 9 (1966): First successful soft landing on the Moon. It proved that spacecraft could land without sinking into dust.

Luna 10 (1966): First spacecraft to orbit the Moon.

Luna 16, 20, 24 (1970s): Sample return missions – brought back lunar soil to Earth.

Lunokhod 1 & 2 (1970–73): Remote-controlled rovers that traveled kilometers across the surface, transmitting soil data and photographs.

### 3. Legacy

The Luna program set the foundation for robotic exploration. Even though Apollo overshadowed it, the Soviet missions achieved many firsts in space history.

## II. India's Chandrayaan Program (2008–Present)

### 1. Chandrayaan-1 (2008)

Launched by ISRO (Indian Space Research Organisation).

Orbited the Moon and created a detailed 3D map of the lunar surface.

Its Moon Impact Probe carried the Indian flag to the Moon.

Discovered water molecules in lunar soil – one of the biggest scientific findings of the decade.

### 2. Chandrayaan-2 (2019)

Consisted of an orbiter, lander (Vikram), and rover (Pragyan).

The orbiter successfully continues to operate, sending high-resolution data.

The lander crash-landed, but ISRO gained crucial technical experience.

### 3. Chandrayaan-3 (2023)

A historic success – India became the first nation to land near the Moon's south pole.

Rover Pragyan confirmed presence of sulfur and other minerals.

Opened new possibilities for future lunar bases, as the south pole contains permanent shadow craters with ice.

### 4. India's Role

Chandrayaan missions placed India in the league of major space powers. They focus on low-cost, high-efficiency exploration, providing global scientific data.

## III. China's Chang'e Program (2007–Present)

Named after the Chinese Moon Goddess Chang'e, this ambitious program is a centerpiece of China's space strategy.

### 1. Early Missions

Chang'e 1 (2007): Orbited and mapped the Moon.

Chang'e 2 (2010): High-resolution mapping and later flew to asteroid Toutatis.

### 2. Breakthrough Achievements

Chang'e 3 (2013): Landed successfully with the Yutu rover.

Chang'e 4 (2019): First-ever landing on the far side of the Moon – a historic global milestone.

Chang'e 5 (2020): Sample-return mission – brought back 2 kilograms of lunar soil.

Chang'e 6 (2024): It was the first lunar mission to retrieve samples from the far side of the Moon; all previous samples were collected from the near side.

### 3. Future Plans

Chang'e 7 & 8 (2026–28): Designed to explore the south pole and test technologies for a permanent lunar research station.

## IV. Comparative Analysis

| Program | Nation | Key Firsts / Achievements |
|---|---|---|
| **Luna** | USSR | First impact, first orbit, first soft landing, first rover, sample return. |
| **Chandrayaan** | India | Discovery of water, first south pole landing, low-cost innovation. |
| **Chang'e** | China | First far-side landing, advanced rovers, modern sample return. |

## V. Importance of Lunar Probes

### Scientific Knowledge

- Surface mapping, mineral analysis, water detection.

### Technology Development

- Rovers, sample return systems, autonomous landing.

### Strategic Goals

- National prestige, future lunar bases, resource mining.

### Human Future

- Ice on the Moon = water, oxygen, rocket fuel.
- These missions pave the way for sustainable human presence.

## Conclusion

The journey from Luna to Chandrayaan to Chang'e shows humanity's collective dream of touching the Moon again and again – not just for glory, but for science and survival.

Each nation has left its footprint:

- The Soviets achieved the firsts.
- India discovered water and polar secrets.
- China mastered modern lunar technology.

Together, they bring us closer to a future where the Moon may become humanity's gateway to the stars.

The first view returned by Luna 3 showed the far side of the Moon was very different from the near side, most noticeably in its lack of lunar maria

**Image:***Chandrayaan-3 – Image of Vikram lander…* by **ISRO / Government of India**, **GODL (Government Open Data License – India)**

**Credit:***Luna-9 Automatic interplanetary station model 1:1* by **Stanislav Kozlovskiy (Wikimedia username: Ctac)**, licensed under **Creative Commons Attribution-ShareAlike 4.0 International (CC BY-SA 4.0)**.

## Why Moon Rocks Matter?

The rocks and soil brought back from the Moon are among the most precious scientific samples in human history.

They not only tell the story of the Moon's geological evolution but also provide crucial insights into the origin of the Earth–Moon system itself.

From Apollo missions (1969–1972) to the Soviet Luna samples, and more recently China's Chang'e-6, each sample is a treasure trove of cosmic history.

## I. Collection of Moon Rocks

### 1. Apollo Missions (1969–1972)

Six Apollo missions successfully returned 382 kilograms of lunar rocks and soil.

Apollo 11 astronauts Neil Armstrong and Buzz Aldrin collected the first 21.5 kg of samples in July 1969.

These samples are carefully stored and preserved at NASA's Johnson Space Center in Houston.

### 2. Soviet Luna Missions

Robotic Luna 16, 20, and 24 brought back 326 grams of lunar soil.

These were the first automated sample return missions in history.

### 3. Recent Contributions

Chang'e-5 (China, 2020): successfully returned 1.7 kg of samples.

Future Missions: India's Chandrayaan-4 (proposed) aim to expand this collection.

## II. Composition of Moon Rocks

Detailed analysis has revealed several distinct types of lunar rocks:

Basalt – Dark, volcanic rock rich in iron and magnesium.

Anorthosite – Bright rocks from the lunar highlands, made mostly of plagioclase feldspar.

Breccia – Mixed rocks formed by meteorite impacts shattering and welding fragments together.

Regolith – The fine dusty soil, created by billions of years of micrometeorite bombardment.

## III. Key Scientific Discoveries

### 1. Age of the Moon

The oldest lunar samples are 4.5 billion years old.

This shows the Moon formed shortly after the Earth itself.

### 2. Giant Impact Hypothesis

Lunar rocks share striking chemical similarities with Earth.

This supports the theory that the Moon was formed when a Mars-sized planet (Theia) collided with early Earth.

### 3. Water on the Moon

Initially, Apollo rocks were believed to be "bone dry."

Later analyses (post-2008) revealed traces of water molecules ($H_2O$, OH groups).

This discovery renewed hopes of future Moon bases utilizing in-situ water.

### 4. Volcanic History

Basalt samples indicate volcanic eruptions occurred on the Moon around 3–4 billion years ago.

Eventually, the Moon lost its internal heat and volcanic activity ceased.

**5. Solar Wind Records**

Tiny particles in the lunar regolith contain trapped solar wind ions.

This provides scientists with a historical record of solar activity over billions of years.

## IV. Distribution and Study

NASA houses the largest collection, but shares portions worldwide.

**"Goodwill Moon Rocks"** – small Apollo samples gifted to many nations as symbols of peace.

Techniques used for study:

- Mass spectrometry
- X-ray diffraction
- Electron microscopy

## V. Importance for the Future

Resource Utilization – Oxygen, metals, and water extraction for human survival.

Planetary Science – Understanding Earth's formation and solar system history.

Human Colonization – Using lunar resources to support long-term settlements.

Space Economy – Future potential for lunar mining industries.

## Conclusion

Moon rocks are not just geological samples – they are time capsules of the early solar system.

They shattered the ancient belief of the Moon as a perfect, unchanging orb and revealed it as a dynamic world.

Every fragment tells a story of billions of years.

And in the future, these rocks may help humanity live permanently on the Moon.

**Panoramic image of the Apollo 17 landing site.**
**NASA**

# 39 - Moon's Water, Volcanic History, And Craters

## From Dry Desert to a Complex World

For decades, scientists believed the Moon was a dry, lifeless world, shaped only by ancient impacts. The Apollo missions reinforced this assumption when early tests on lunar rocks showed no trace of water. But in the last two decades, new technologies and probes have completely rewritten the Moon's story.

We now know the Moon is not entirely dry. It has traces of water ice in shadowed craters, volcanic activity in its past, and a surface scarred by billions of years of cosmic collisions.

This chapter explores the Moon's water, volcanic history, and craters – three elements that define its true nature.

## I. Moon's Water – The Hidden Treasure

### 1. Early Belief: A Dry Moon

Apollo samples (1969–1972) were initially thought to be bone dry.

The harsh vacuum of space and lack of atmosphere suggested water couldn't survive on the surface.

### 2. Discovery of Water

India's Chandrayaan-1 mission (2008) confirmed water molecules ($H_2O$ and OH) spread across the surface.

NASA's LCROSS mission (2009) detected large amounts of water ice in permanently shadowed craters near the lunar poles.

### 3. Sources of Lunar Water

Comets and Asteroids: Delivered water through impacts.

Solar Wind: Hydrogen ions from the Sun combine with oxygen in lunar soil to form water molecules.

Internal Sources: Volcanic eruptions may have brought water from the lunar interior.

### 4. Importance of Water

Future human colonies can use lunar water for drinking, oxygen production, and rocket fuel (via hydrogen and oxygen separation).

Water ice makes the Moon a potential gateway for space exploration.

## II. Volcanic History of the Moon

### 1. Volcanic Plains – The Lunar Maria

The dark patches visible from Earth, called maria (Latin for "seas"), are vast basalt plains formed by ancient lava flows.

These cover about 16% of the Moon's surface, mainly on the near side.

### 2. Timeline of Volcanic Activity

Active volcanic eruptions occurred between 3–4 billion years ago.

The youngest lava flows may be only 100 million years old, suggesting the Moon remained geologically active for longer than once believed.

### 3. Types of Volcanic Features

Shield Volcanoes: Gentle slopes formed by low-viscosity lava.

Rilles (Lava Channels): Long, winding valleys created by flowing lava.

Domes: Rounded hills formed by thicker lava pushing upward.

Volcanic Glass Beads: Tiny spheres of cooled lava, some containing water molecules.

### 4. Why Volcanism Stopped

The Moon's small size meant it lost heat quickly.

Without internal heat, volcanic activity eventually ceased.

## ✍III. Craters – Scars of Celestial Violence

### 1. Formation of Craters

Caused by meteorite and asteroid impacts.

With no atmosphere, the Moon has no protection, so impacts have accumulated over billions of years.

### 2. Major Craters and Basins

Tycho Crater: 85 km wide, famous for its bright rays.

Copernicus Crater: 93 km wide, a classic example of a young impact crater.

South Pole–Aitken Basin: One of the largest impact structures in the solar system, about 2,500 km wide.

### 3. Crater Features

Central Peaks: Caused when the surface rebounds after impact.

Ejecta Blankets: Rock and dust thrown out around the crater.

Ray Systems: Bright streaks radiating from fresh craters.

### 4. Scientific Importance of Craters

They preserve a record of the solar system's violent history.

Impact melt and debris help estimate the ages of lunar regions.

Some craters act as cold traps for water ice.

## IV. Philosophical and Scientific Implications

From Dry Rock to Resourceful World: The discovery of water completely shifted human perspectives on lunar colonization.

Volcanism and Planetary Evolution: Studying lunar volcanism reveals how smaller bodies cool and evolve compared to Earth.

Craters as Cosmic History Books: The Moon serves as a natural archive, preserving impact history erased on Earth by weather and tectonics.

## Conclusion

The Moon is not a static, lifeless rock.

Its water ice hints at sustainability for future explorers.

Its volcanic plains reveal a fiery, dynamic past.

Its craters bear witness to billions of years of cosmic violence.

Together, these features make the Moon not just Earth's companion, but a scientific key to understanding planetary evolution and the origins of our solar system.

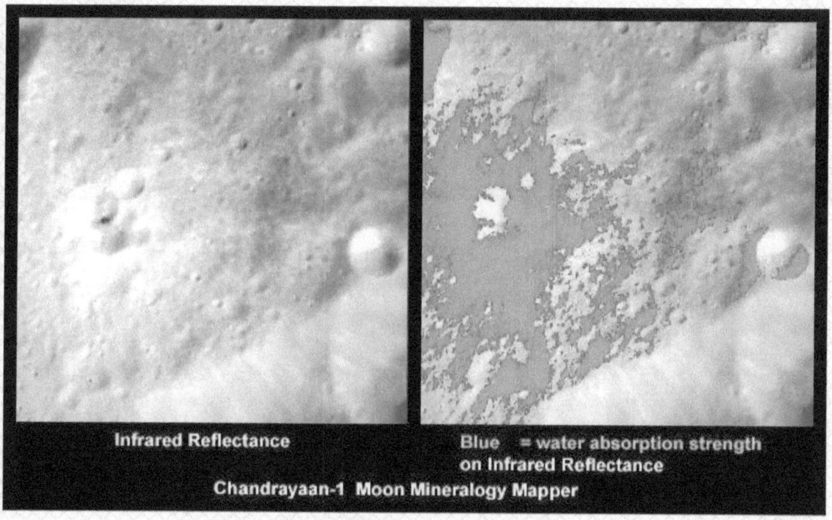

**"Imbrium location"** by Srbauer, licensed under **CC BY-SA 3.0** — used withcredit.*Source:* Wikimedia Commons

# 40 - International Lunar Missions And Collaborations

## From National Rivalries to Global Partnerships

The 20th century "Space Race" between the United States and the Soviet Union was largely a competition—each nation striving to prove its scientific, political, and military supremacy. But in the 21st century, the focus of space exploration has shifted dramatically.

The Moon is no longer a symbol of rivalry but a platform for cooperation. With the discovery of lunar water, renewed interest in sustainable human exploration, and plans for lunar bases, countries are increasingly pooling their resources, technologies, and scientific knowledge.

Today, lunar exploration is not just NASA or Roscosmos; it includes Europe, India, China, Japan, South Korea, Israel, the UAE, and private companies like SpaceX, Blue Origin, and Astrobotic. Together, these efforts represent the dawn of a truly international and multi-stakeholder era of Moon exploration.

## I. Europe and the ESA (European Space Agency)

The European Space Agency (ESA), with 22 member states, has contributed significantly to lunar exploration through partnerships and independent missions.

**SMART-1 (2003–2006):**

Europe's first mission to the Moon. It tested solar-electric propulsion and performed mineralogical surveys of the lunar surface.

**Collaboration with NASA Artemis:**

ESA is a key partner in Artemis, providing the European Service Module (ESM) for the Orion spacecraft—crucial for human missions to the Moon and beyond.

**Future Goals:**

HERACLES: A robotic mission planned for 2028 to test human-robot interaction and bring lunar samples back to Earth from orbit.

Argonaut: A new program for European robotic landers, with the first mission targeting the Moon's south pole around 2030, aiming to deliver payloads and demonstrate landing tech.

Moonlight: Establishing a lunar satellite constellation for communication and navigation, similar to GPS, to support future missions.

ESA envisions a "Moon Village" – a concept where different nations share infrastructure on the Moon for science, commerce, and resource use.

## II. China's Chang'e Program

China's Chang'e program has been one of the most ambitious and successful modern lunar efforts.

Chang'e-3 (2013): Landed Yutu rover – China's first rover.

Chang'e-4 (2019): First mission to land on the far side of the Moon – a historic milestone.

Chang'e-5 (2020) & 6(2024): Successfully returned 2 kg of lunar samples to Earth – first sample return since the 1970s.

Future Plans:

Chang'e-7 and 8: Testing technologies for lunar base construction.

By 2030s: Chinese crewed lunar landing.

China collaborates with Russia on the proposed International Lunar Research Station (ILRS), aiming for the 2030s.

## III. India's Chandrayaan Program

India, through ISRO (Indian Space Research Organisation), has become a rising lunar power.

Chandrayaan-1 (2008):

Discovered water molecules on the Moon—a groundbreaking contribution.

Chandrayaan-2 (2019):

Orbiter successful; lander crashed, but orbiter continues valuable science.

Chandrayaan-3 (2023):

Historic success—India became the fourth nation to soft-land on the Moon and the first at the lunar south pole.

ISRO is in talks for international collaborations, including NASA-ISRO partnerships for future missions.

## IV. Japan's JAXA Missions

Japan has a long history of lunar missions under JAXA (Japan Aerospace Exploration Agency).

SELENE / Kaguya (2007–2009): A large lunar orbiter that mapped gravity, topography, and mineralogy.

OMOTENASHI (2022, part of Artemis I): A tiny CubeSat lander (attempt unsuccessful).

SLIM (2024): The "Moon Sniper" mission, aiming for pinpoint landing accuracy within 100 meters.

Japan collaborates with NASA on Artemis and contributes to Gateway, the planned lunar space station.

## V. South Korea – KPLO (Danuri)

Korea Pathfinder Lunar Orbiter (Danuri, 2022):

South Korea's first lunar mission, launched with help from NASA and SpaceX.

It studies the Moon's magnetic fields, resources, and landing site candidates for future missions.

### VI. Russia – Luna Program Revived

The Soviet Union once dominated lunar exploration with its Luna program. In the modern era:

Luna-25 (2023): Attempted south pole landing but crashed.

Future Plans: Luna-26 and Luna-27 for orbital mapping and lander missions.

Collaborating with China on ILRS (International Lunar Research Station).

### VII. USA – From Apollo to Artemis

After the Apollo era, NASA paused lunar landings, but interest has surged again.

Artemis Program (2020s–2030s):

Goal – return humans to the Moon by 2025–26.

Artemis I (2022): Successful uncrewed Orion test.

Artemis II (2026): Planned crewed lunar orbit.

Artemis III (2027): Planned crewed lunar landing at the south pole.

Commercial Lunar Payload Services (CLPS):

Partnering with companies like Astrobotic and Intuitive Machines to send robotic landers.

### VIII. International Collaborations

**1. Artemis Accords**

A U.S.-led framework for responsible lunar exploration.

Over 30+ countries signed (including India, Japan, Canada, EU nations).

Promotes transparency, peaceful use, and sharing of data.

**2. International Lunar Research Station (ILRS)**

Proposed by China and Russia.

Aimed to rival Artemis, focusing on shared bases and science by the 2030s.

**3. Private Sector & Global Partnerships**

Companies (SpaceX, Blue Origin, ispace) are key players.

UAE, Luxembourg, and others fund private lunar startups.

## IX. The Future – Toward a Permanent Human Presence

By the 2030s–2040s, we may see:

Permanent Moon bases (Artemis Gateway + ILRS).

Large-scale mining of lunar water for rocket fuel.

Tourism and commercial outposts.

Scientific "Moon villages" involving dozens of nations.

The Moon is emerging as a test-bed for Mars exploration and a hub for the new space economy.

## Conclusion

From rivalry to collaboration, the Moon has become a global stage for cooperation.

The U.S. leads with Artemis, joined by allies.

China and Russia push ILRS as an alternative.

India, Japan, Korea, Europe, and others contribute unique strengths.

Private companies expand opportunities beyond governments.

The story of lunar exploration is no longer about one flag or one nation, but about the entire human race reaching for the stars together.

NASA : A render of the Artemis Base Camp

render of the ILRS mission

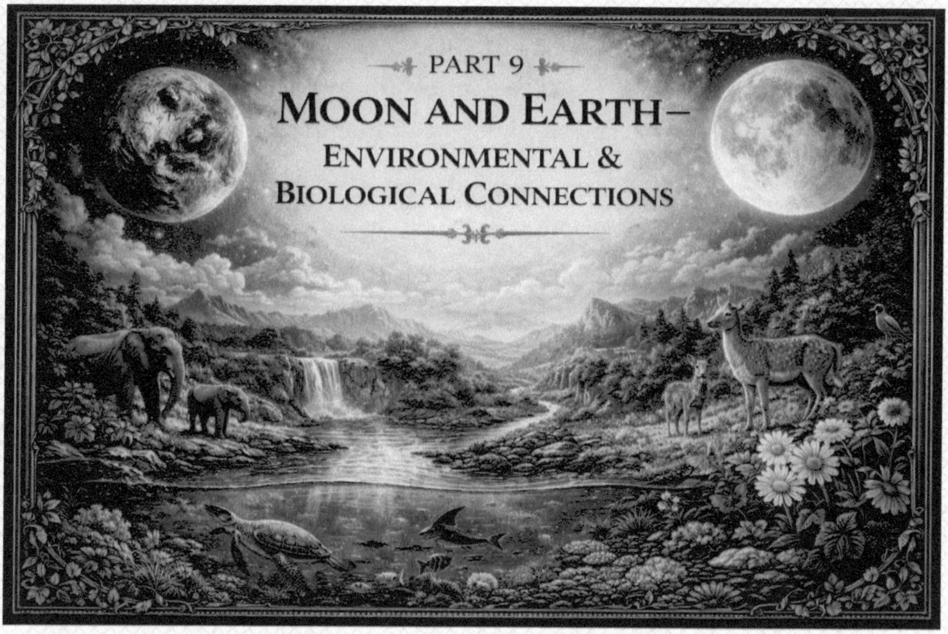
PART 9
MOON AND EARTH–
ENVIRONMENTAL &
BIOLOGICAL CONNECTIONS

# 41- Tidal Energy And Ocean Ecology

## The Power of the Moon and the Ocean

For thousands of years, humans have observed how the Moon controls the tides. The rhythmic rise and fall of the ocean, caused by the gravitational pull of the Moon (and partly the Sun), has shaped coastal life, fishing practices, navigation, and cultural beliefs.

In modern times, tides have taken on a new dimension – a renewable source of clean energy. At the same time, tidal movements deeply influence marine ecosystems, biodiversity, and climate regulation.

This chapter explores how tidal forces create energy potential, their role in ocean ecology, and the balance required between human development and nature.

## I. Science of Tides

Tides occur due to the gravitational interaction between the Moon, Earth, and Sun.

High tide happens when ocean water bulges toward the Moon.

Low tide occurs in areas at right angles to this bulge.

The Sun reinforces or weakens tides depending on alignment:

Spring Tides (New Moon/Full Moon) – stronger tides.

Neap Tides (Quarter Moon) – weaker tides.

Tidal cycles are predictable, which makes them an attractive source of renewable energy.

## ⚡ II. Tidal Energy – Harnessing Lunar Power

Tidal energy is one of the most reliable forms of renewable energy because tides follow predictable astronomical cycles.

### 1. Tidal Range Systems (Barrage Method)

Dams or barrages are built across estuaries.

During high tide, water enters the basin; during low tide, it flows out through turbines.

Example: La Rance Tidal Power Station, France (operational since 1966).

### 2. Tidal Stream Systems

Turbines are placed underwater in strong tidal currents.

Works like underwater wind turbines.

Example: MeyGen Project, Scotland – world's largest tidal stream project.

### 3. Dynamic Tidal Power (Proposed)

Large dams extending into the sea to capture differences in tidal phases.

### 4. Advantages of Tidal Energy

Predictable and consistent.

Zero carbon emissions once installed.

Long operational life.

### 5. Challenges

High construction costs.

Environmental disruption (estuaries, fish migration).

Limited suitable sites globally.

## III. Ecological Role of Tides in Ocean Life

Tides are not just about energy; they are essential to ocean ecology.

### 1. Nutrient Cycling

Tidal flows carry nutrients between deep and shallow waters.

Estuaries (where rivers meet the sea) are nutrient-rich nurseries for fish and crustaceans.

**2. Coastal Habitats**

Mangroves, mudflats, and salt marshes depend on tidal flooding.

These habitats provide shelter for migratory birds, crabs, and juvenile fish.

**3. Marine Biodiversity**

Intertidal zones (between high and low tide marks) are among the most biodiverse ecosystems on Earth.

Creatures like barnacles, mussels, and starfish adapt to alternating wet and dry conditions.

**4. Ocean Circulation**

Tidal forces mix coastal waters, affecting temperature, salinity, and oxygen levels.

This mixing regulates local climates and sustains fisheries.

## IV. Human Dependency and Ocean Cultures

**Historically, tides dictated:**

- Fishing cycles – Fishermen still follow tides to know when fish are abundant.
- Navigation and trade – Ancient sailors planned voyages based on tidal timings.
- Cultural practices – Many coastal cultures hold rituals during full-moon high tides (e.g., harvest festivals, sea worship).

## ⚖V. Environmental Concerns of Tidal Power

While tidal energy is clean, it can disturb marine ecosystems if not carefully planned.

Barrage projects may block fish migration routes (e.g., salmon).

Sediment flow disruption can alter coastal erosion patterns.

Underwater turbines may affect marine mammals (dolphins, whales).

Noise pollution may interfere with echolocation of aquatic species.

Modern projects use eco-friendly designs – fish-friendly turbines, open barrages, and careful site selection.

## VI. Global Case Studies

France – La Rance Barrage

Produces 240 MW power.

Initially caused ecological disruption but later stabilized into a new estuarine ecosystem.

South Korea – Sihwa Lake Tidal Power Station

World's largest (254 MW).

Helped restore water quality in the previously polluted lake.

United Kingdom – MeyGen Project (Scotland)

Tidal stream project with potential of 400 MW.

Focus on minimal ecological disruption.

India – Gulf of Kutch and Gulf of Khambhat

Huge tidal potential identified.

Projects under consideration for renewable energy needs.

## VII. The Future – Balance of Energy and Ecology

The future of tidal energy depends on technological innovation and ecological sensitivity.

Hybrid models combining tidal + wind + solar.

Floating tidal turbines to reduce seabed disruption.

International policies ensuring marine conservation.

At the same time, climate change is raising sea levels – altering tidal patterns and threatening coastal ecosystems. Understanding the Moon's influence on tides is now critical for climate adaptation.

## Conclusion

The Moon's pull on Earth does more than move water – it shapes human civilization and ocean life.

Tidal Energy represents humanity's attempt to transform lunar gravity into sustainable electricity.

Ocean Ecology shows how tides nurture biodiversity and regulate climate.

The challenge is to balance technological progress with ecological preservation.

As humanity moves toward a renewable future, the tides – driven by the Moon - may become one of the most powerful allies of sustainable living on Earth.

# 42- Animal Behavior During Lunar Phases

## Nature's Clock in the Sky

The Moon has always served as a cosmic timekeeper, not just for humans but also for animals. From the deepest oceans to dense forests and open savannas, countless species synchronize their behavior with lunar phases.

Scientists call this lunar biology or moonlight ecology – a field that studies how moonlight, gravity, and lunar rhythms affect living creatures. While the Sun rules circadian rhythms (daily cycles), the Moon influences circalunar rhythms (monthly cycles).

This chapter explores how different animals – from fish and birds to insects and mammals – respond to the waxing and waning Moon.

## I. The Science of Lunar Influence

Moonlight intensity – Full Moon nights are bright, affecting nocturnal animals' visibility.

Tides and gravity – Marine animals use tidal rhythms (linked to the Moon) for feeding and breeding.

Predator-prey dynamics – Moonlight affects hunting efficiency.

Reproductive cycles – Many species synchronize mating with specific lunar phases.

## II. Marine Life and Lunar Cycles

### 1. Coral Spawning

Tropical corals release eggs and sperm in a mass synchronized event during Full Moon nights.

Example: Great Barrier Reef, Australia – hundreds of coral species coordinate spawning within a few nights.

### 2. Fish Behavior

Grunion fish (California) spawn on sandy beaches only during spring tides around Full Moon.

Reef fish plan mating and egg-laying to maximize survival, often guided by lunar light.

### 3. Crustaceans and Mollusks

Crabs migrate to shorelines for egg release during certain lunar phases.

Squid are known to adjust hunting behavior based on moonlight.

## III. Birds and Lunar Guidance

Migratory birds sometimes use the Moon and stars for navigation.

Owls and other nocturnal birds adjust their hunting on bright nights, avoiding overexposure to predators.

Certain seabirds (like petrels) return to nests more frequently on darker nights (New Moon).

## IV. Insects and Lunar Cues

Moths navigate using the Moon; artificial lights confuse them (light pollution problem).

Fireflies alter flashing behavior during Full Moon to avoid predators.

Mosquito activity often peaks around New Moon when nights are darker.

## V. Mammals and the Full Moon Effect

### 1. Predators

Lions hunt less during Full Moon since prey can see them more easily.

Wolves, often symbolically linked with the Moon, are not proven to howl more at Full Moon, but they do use moonlit nights for coordinated group hunting.

### 2. Prey Animals

Deer, rabbits, and rodents reduce movement on Full Moon nights to avoid predators.

Studies show higher heart rates and vigilance in small mammals during moonlit nights.

### 3. Humans and Folklore

Historically, people linked lunacy, sleeplessness, and crime rates to Full Moons.

Scientific studies show some correlation between lunar phases and human sleep cycles, but evidence on behavior is mixed.

## VI. Case Studies

### 1. Sea Turtle Nesting

Female turtles come ashore during New Moon high tides to lay eggs, minimizing risk of predation.

Hatchlings often emerge during darker nights, reducing visibility to predators like birds.

### 2. Horseshoe Crabs

Along the U.S. East Coast, horseshoe crabs lay eggs on beaches during Full and New Moon tides.

This event feeds thousands of migratory shorebirds, creating a vital ecological chain.

### 3. Bats and Moonlight

Many bat species reduce activity during Full Moon – a phenomenon called lunar phobia – to avoid being spotted by predators.

## VII. The Ecological Balance

Predator-prey balance – Moonlight creates cycles of visibility and hiding.

Reproductive success – Lunar timing increases chances of offspring survival.

Ecosystem synchronization – Entire food chains (like crab-egg–shorebird cycle) depend on lunar phases.

## VIII. Future Research – Moonlight in a Changing World

Light pollution – Artificial lighting disrupts natural lunar cues for animals.

Climate change – Rising sea levels may disturb tide-based breeding.

Space exploration link – Understanding lunar biology may help in predicting how life could adapt to extraterrestrial moons.

## Conclusion

The Moon silently orchestrates life on Earth. From the spawning of corals to the cautious steps of deer, from guiding migrating birds to triggering sea turtle nesting, lunar phases are an invisible rhythm of nature.

The study of animal behavior during lunar phases reveals a deeper truth: Earth's life is inseparable from cosmic forces. The Moon is not just a rock in the sky – it is a living clock, shaping evolution, behavior, and survival on our planet.

Horseshoe crabs spawn high tide Slaughter Beach Delaware Bay waves

Image: NOAA (Public Domain)

# 43- Human Sleep, Psychology, And The Moon

## The Eternal Connection Between Moon and Mind

From the dawn of human civilization, the Moon has been more than a celestial body in the night sky. It has been a mystical guide, a psychological influence, and a regulator of human rhythms.

While modern science often relies on biological clocks tied to the Sun (circadian rhythms), research increasingly suggests that the Moon and its phases also shape human sleep, mood, and behavior.

Ancient cultures associated lunacy (madness) with the Full Moon.

Modern studies reveal that sleep cycles shorten around the Full Moon.

Psychologists note increased dream activity, anxiety, and emotional shifts tied to lunar phases.

This chapter explores in depth how the Moon impacts human sleep patterns, psychological states, mental health, dreams, creativity, and even psychiatric disorders. We'll weave together history, myths, scientific studies, case reports, and modern neuroscience to present a complete picture.

## I. Biological Rhythms: Beyond the Circadian Clock

### 1. Circadian vs. Circalunar Rhythms

Circadian rhythm (24h) → controlled by light/dark cycle of the Sun, regulated by melatonin.

Circalunar rhythm (~29.5 days) → aligned with the Moon's phases.

Many living beings (including humans) show circalunar biological responses – even when isolated from natural light. This suggests the body carries an internal lunar clock.

### 2. Evolutionary Basis

Early humans evolved under bright Full Moons (useful for hunting, travel, rituals).

New Moons created darkness – influencing safety, reproduction, and behavior.

These survival patterns may still shape our subconscious biological responses.

## II. Sleep and the Moon

### 1. Scientific Evidence

Several controlled experiments show:

- During Full Moon, humans sleep 20–30 minutes less.
- REM sleep (dreaming stage) reduces, but dream recall may increase.
- Melatonin secretion drops around Full Moon nights.

Example: A 2013 University of Basel study (Cajochen et al.) found:

- Participants slept shorter and lighter during Full Moon.
- They had lower melatonin levels.
- Reported less restful sleep.

### 2. Possible Mechanisms

Moonlight exposure: even subtle brightness at night disrupts melatonin.

Gravitational pull: debated, but some scientists suggest biological fluids may respond to lunar tides.

Psychological expectation: awareness of Full Moon may cause subconscious sleep disturbance.

### 3. Cultural Impact on Sleep

In villages without artificial lights, Full Moon nights often extend waking hours.

Rituals, festivals, or fear-based taboos affect sleep behavior.

## III. Psychological Influence of the Moon

### 1. Emotional States

Full Moon → heightened energy, anxiety, irritability in some people.

New Moon → feelings of low energy, introspection.

### 2. "Lunacy" and History

Term "lunatic" comes from Latin lunaticus, meaning "moonstruck."

Medieval Europe believed Full Moons triggered madness, violence, and epilepsy.

Even today, emergency rooms and police departments sometimes report higher incidents during Full Moon (though scientific evidence is mixed).

### 3. Modern Psychiatry

Some psychiatric disorders appear sensitive to lunar cycles:

Bipolar disorder – mood swings may worsen during Full Moon.

Epilepsy – seizures sometimes reported more frequently in certain lunar phases.

Sleepwalking – occasionally linked to lunar cycles.

## IV. Dreams and the Moon

### 1. Dream Intensity

People often report more vivid, strange, or unsettling dreams during Full Moon.

Reduced REM duration may compress dream cycles, intensifying memory recall.

### 2. Archetypes in Dream Psychology

Moon is a universal archetype in Jungian psychology: symbolizing femininity, mystery, and unconscious mind.

Lunar dreams often involve water, tides, transformation, and hidden fears.

**3. Case Studies**

Ancient shamans timed dream rituals with lunar phases.

In some Native American tribes, Full Moon dreams were considered prophetic.

## V. Creativity, Art, and the Moon's Inspiration

**1. Lunar Muse**

The Moon has inspired countless works of art, poetry, and music (from Shakespeare's sonnets to Beethoven's Moonlight Sonata).

**2. Psychological Basis**

Increased wakefulness during Full Moon → more creative output at night.

Altered sleep → access to hypnagogic states (between sleep and wakefulness), known to boost creativity.

**3. Artists and Writers**

Vincent van Gogh's "Starry Night" painted under influence of moonlit skies.

Romantic poets (Shelley, Byron, Tagore) often tied inspiration to lunar imagery.

## ⚖VI. Myth vs Science: Criminal Behavior & Moon

**1. Folklore Beliefs**

Werewolves, vampires, witches – all tied to Full Moon.

Police folklore: "more crimes happen on Full Moon nights."

### 2. Research Findings

Large studies find no consistent rise in crime or hospital admissions.

But some evidence suggests slight increases in traffic accidents, psychiatric emergencies, or ER visits.

### 3. Psychology of Belief

People expect unusual things to happen on Full Moon → confirmation bias makes them notice more events.

## VII. Women, Fertility, and the Moon

### 1. Menstrual Cycles

Average cycle (29.5 days) nearly matches lunar month.

Historically, women ovulated around Full Moon ("white moon cycle") or New Moon ("red moon cycle").

### 2. Fertility Rituals

Many cultures tied fertility rituals to lunar phases.

Example: Ancient Greeks associated Artemis (moon goddess) with childbirth.

### 3. Modern Studies

Some women still report menstrual synchronization with lunar phases.

Artificial light may disrupt this natural rhythm.

## VIII. Neuroscience and the Moon

### 1. Brain and Melatonin

Pineal gland produces melatonin in darkness.

Lunar brightness disrupts melatonin → alters circadian clock.

### 2. Magnetoreception & Tides

Theories suggest subtle gravitational pull could affect brain fluid dynamics.

Still controversial, but animal studies hint at magneto-sensitivity.

**3. Placebo and Awareness**

Psychological anticipation ("It's a Full Moon, I won't sleep") may itself reduce sleep.

## IX. Case Studies Around the World

India: People still fast or meditate on Full Moon (Purnima) → psychological calmness despite less sleep.

Japan: Moon-viewing festivals (Tsukimi) combine aesthetics with emotional reflection.

Africa: Hunters use Full Moon for extended night hunts.

Western countries: Reports of ER staff bracing for "Full Moon madness."

## X. Modern Applications

**1. Sleep Research**

Understanding lunar influences may help design better sleep therapies.

**2. Psychiatry**

Tracking mood changes with lunar cycles may help manage bipolar or anxiety disorders.

**3. Space Exploration**

Astronaut sleep cycles on Moon missions may require considering lunar light exposure.

## Conclusion – The Moon as the Mirror of Mind

The Moon's pull on the Earth shapes tides, oceans, and weather – but it also seems to reach into the human mind, dreams, and emotions.

While not every folklore claim is scientifically true, growing evidence shows:

The Moon subtly affects sleep duration, dream intensity, and emotional balance.

Our psychological connection is as real as our poetic imagination.

Thus, the Moon is both a scientific regulator and a psychological archetype – shaping not only how we sleep but also how we dream, create, and understand ourselves.

# 44 - Moon's Role In Earthquake And Weather Theories

## A Celestial Body with Terrestrial Effects

Since ancient times, humans have speculated about whether the Moon does more than control the tides. Could its gravitational pull or phases also influence earthquakes, storms, rainfall, or even seasonal shifts?

Farmers once planned sowing and harvesting by lunar cycles.

Sailors relied on Full Moon tides for navigation.

Priests and astrologers linked lunar alignments to droughts, floods, or quakes.

In modern times, geophysicists, seismologists, and meteorologists have re-examined these questions. While scientific evidence is mixed, the Moon's gravity, light, and orbital dynamics undeniably interact with Earth's geophysical systems. This chapter explores myths, ancient theories, modern studies, and ongoing debates about the Moon's connection to earthquakes and weather.

## I. Gravitational Influence of the Moon

1. Tides and Ocean Dynamics

The Moon's gravity pulls ocean water → creating high and low tides.

Spring tides (Full & New Moon) → higher than normal.

Neap tides (Quarter Moon) → lower than normal.

This tidal force not only shapes marine ecosystems but also exerts stress on Earth's crust.

2. Earth Tides (Crustal Movements)

The solid Earth itself experiences "Earth tides" – microscopic rises and falls of a few centimeters due to lunar gravity.

These subtle stresses can, in theory, affect fault lines already close to rupture.

## II. The Moon and Earthquakes

### 1. Ancient Beliefs

In China, earthquakes were sometimes blamed on "Heavenly alignments" including lunar phases.

Japanese folklore linked tsunamis and earthquakes to "angry sea gods" rising during Full Moons.

Indian texts occasionally associated amavasya / poornima with calamities.

### 2. Scientific Studies

Modern Seismology:

Some research shows small statistical increases in earthquake frequency during Full or New Moon, when tidal stresses peak.

However, most earthquakes occur randomly and cannot be predicted by lunar phase alone.

2016 – Nature Geoscience Study:

Found that large subduction zone quakes (e.g., Japan, Chile) may be slightly more likely during Spring Tides.

Reason: stronger gravitational stress adds to tectonic pressure.

### 3. Ongoing Debate

Majority view → Moon cannot directly cause quakes, but may trigger already primed faults.

Minority view → Lunar and solar gravitational alignments could play a small but real role in earthquake timing.

## ♣III. The Moon and Weather Patterns

### 1. Historical Agricultural Theories

Lunar farming calendars in Europe and India → advised sowing seeds on waxing moon, harvesting on waning moon.

Belief: Moon influenced soil moisture, rainfall, and crop yield.

### 2. Rainfall and Lunar Cycles

Some studies suggest slight increases in rainfall around Full Moon.

Hypothesis: Lunar gravity influences atmospheric tides → altering cloud formation.

NASA has confirmed "atmospheric tides" caused by Moon's pull, but their effect on rainfall is still debated.

### 3. Cyclones and Storms

Folklore: "Storms rage stronger under the Full Moon."

Science: While storm intensity is driven mostly by ocean heat, some models suggest tidal stresses may amplify ocean-atmosphere coupling.

## IV. Theories from the Middle Ages to Modern Science

1. Medieval Theories

Astrologers believed lunar eclipses predicted floods, plagues, or earthquakes.

Physicians tied illness and fevers to lunar weather cycles.

2. 19th & 20th Century Science

Early meteorologists attempted to correlate lunar phases with rainfall.

Mixed results → some patterns observed, but inconsistent.

3. Modern Satellite Observations

Satellite data shows Moon does indeed affect atmospheric pressure waves.

But direct impact on daily weather remains small compared to solar energy.

## V. Case Studies

1. The 2004 Indian Ocean Earthquake & Tsunami

Occurred just after Full Moon → some speculated lunar tides played a role.

Later analysis showed tectonic stress was the primary cause.

2. Hurricanes in the Atlantic

Some hurricanes peaked near Full Moons.

Scientists remain cautious, noting correlation ≠ causation.

3. Volcanic Eruptions

Studies of volcanoes in Hawaii & Iceland → eruptions sometimes coincided with Spring Tides.

Lunar stress may "tip the balance" in magma chambers near bursting point.

## VI. Current Research Directions

**Seismology:**

Using lunar cycles to model earthquake risk in high-stress zones.

**Climate Science:**

Studying atmospheric tides caused by the Moon's pull.

Linking them to monsoons, rainfall variability, and jet streams.

**Space Science:**

Understanding Earth-Moon gravitational system helps in predicting planetary climate models for exoplanets.

## ⚖ VII. Myth vs Reality

✅ Reality: Moon controls tides, atmospheric pressure, and may trigger quakes already on the edge.

❌ Myth: Moon singlehandedly causes disasters, floods, or quakes.

✅ Reality: Subtle correlations exist, but major weather/climate is solar-driven.

❌ Myth: Every Full Moon brings chaos – mostly cultural belief.

## Conclusion – The Moon as a Subtle Earth-Shaper

The Moon is not only a night lamp or a tide maker but also a subtle influencer of Earth's geophysical balance.

It stresses the crust → sometimes nudging faults to slip.

It shapes ocean and atmospheric tides → perhaps modulating rainfall.

It inspires both folklore fears and scientific curiosity.

In the end, the Moon's role in earthquakes and weather remains a fascinating blend of myth, partial evidence, and ongoing science. While it may not command storms or quakes, it is an invisible participant in Earth's dynamic systems – a reminder that our planet and its satellite are locked in a deep, evolving dance.

# 45 - Lunar Farming And Agricultural Practices

## The Moon as a Farmer's Calendar

For thousands of years, farmers across civilizations have looked to the Moon as a natural calendar and guide for planting, harvesting, and livestock care. Long before scientific instruments and meteorological forecasts, the Moon's phases and its gravitational pull on Earth's waters shaped the rhythm of agricultural life.

Ancient Egyptians, Mesopotamians, and Mayans scheduled sowing according to lunar phases.

In India, Vedic texts mention Nakshatras and lunar timings for planting grains.

In Europe, the "Farmer's Almanac" carried lunar planting guides for centuries.

Indigenous communities worldwide still rely on moon-based farming traditions.

Today, lunar agriculture is studied under biodynamic farming and ecological agriculture, where cosmic and lunar cycles are believed to enhance plant vitality.

## I. The Science and Belief Behind Lunar Agriculture

### 1. Gravitational Influence

Just as the Moon controls tides, it may subtly affect soil moisture.

Seeds planted during waxing moon phases reportedly absorb more water.

Farmers noticed quicker germination during certain lunar phases.

### 2. Light Influence

Brightness of the Moon affects nocturnal insects and pollinators.

Increased light around the Full Moon encourages certain crops to sprout better.

**3. Myth vs Science**

Modern science remains divided: controlled experiments show mixed results.

Yet, millions of farmers worldwide swear by lunar calendars.

## II. Phases of the Moon and Planting Cycles

| Moon Phase | Astronomical Meaning | Traditional Farming Belief | Suggested Crops |
|---|---|---|---|
| New Moon | Moon invisible | Seeds absorb water, good for germination | Leafy greens, grains |
| Waxing Crescent | Moonlight increases | Upward energy, stronger above-ground growth | Tomatoes, spinach, herbs |
| Full Moon | Brightest, max gravity | Strongest pull on water, best for flowering & fruiting | Cucumbers, melons, beans |
| Waning Moon | Light decreases | Energy moves downwards, strengthens roots | Root crops – carrots, potatoes, onions |

## III. Biodynamic Agriculture and Lunar Cycles

Biodynamic farming, pioneered by Rudolf Steiner in the early 20th century, goes beyond organic practices by using lunar and cosmic rhythms.

Uses a lunar sowing calendar.

Divides crops by elemental association:

Root (Earth) → planted in waning phases.

Leaf (Water) → planted in waxing phases.

Flower (Air) → near Full Moon.

Fruit (Fire) → just before Full Moon.

Farmers also prepare special composts energized by moon phases, believing it enhances soil fertility.

## IV. Cultural Perspectives

1. India

Purnima (Full Moon) and Amavasya (New Moon) play roles in agricultural rituals.

Farmers sow paddy during waxing moon for abundant yield.

2. China

The Chinese lunar calendar guides planting festivals.

Certain crops like rice are sown after spring lunar festivals.

3. Europe

Medieval farmers relied on "lunar almanacs."

"Plant potatoes on a waning moon" is still common folklore.

4. Mayan Civilization

Mayans developed highly detailed lunar calendars to schedule maize cultivation.

## V. Livestock and Lunar Cycles

Farmers note that animals show more restless behavior during Full Moon nights.

Shearing sheep and milking cows were sometimes timed with lunar phases.

Folk wisdom: eggs hatched under waxing moon grow healthier chickens.

## VI. Scientific Studies and Modern Relevance

Swiss Research (2013): Certain plants sprouted faster during waxing moon.

Brazilian Coffee Farmers: Report higher yields when planting aligned with lunar cycles.

NASA Experiments: No conclusive lunar effect in controlled greenhouses.

Modern application:

Many organic farmers integrate lunar calendars as an additional tool.

Even if partially symbolic, it helps farmers stay mindful of natural cycles.

### Conclusion – The Moon as Nature's Partner

The Moon remains a timeless companion to agriculture. Whether through gravitational pull, moonlight, or human cultural memory, it continues to inspire farming practices.

For some, it is science + tradition.

For others, it is symbolic harmony with nature.

Regardless, lunar agriculture highlights humanity's bond with cosmic rhythms.

"Farm with the Sun for the day, but trust the Moon for the season."

**indian farmers tying seeds in cloth and keeping them under moonlight before sowing.**

PART X
FUTURE OF THE MOON
COLONIES, MINING &
MYSTERIES

# 46 - Moon Bases And Lunar Tourism

**From Imagination to Reality**

For centuries, poets, dreamers, and scientists have looked up at the Moon and envisioned a day when humans might live there. What was once confined to science fiction novels and artistic sketches is now actively being discussed in space agencies, private companies, and global collaborations.

Moon bases – permanent or semi-permanent human settlements on the lunar surface – are no longer just fantasy. With advancements in spaceflight technology, robotics, 3D printing, and sustainable life-support systems, humanity is closer than ever to making the Moon a new home. Alongside science, a new industry is emerging: Lunar Tourism – the idea of sending ordinary (though very wealthy) travelers to orbit, land, and even vacation on the Moon.

This chapter explores the history, current plans, challenges, and future prospects of Moon bases and lunar tourism, weaving together technology, economics, politics, and human imagination.

## I. Historical Roots of Lunar Settlement Ideas

### 1. Early Concepts

1600s–1800s: Philosophers like Johannes Kepler and later authors speculated about life on the Moon.

1865: Jules Verne's From the Earth to the Moon depicted the first fictional lunar voyage.

1902: Georges Méliès' silent film A Trip to the Moon showed people exploring lunar landscapes.

### 2. The Space Race Era

After Apollo landings (1969–1972), NASA considered building a lunar base by the 1980s, but funding and politics halted progress.

The Soviet Union also imagined permanent lunar stations, but collapse of the USSR ended those ambitions.

3. The Modern Revival

In the 21st century, new interest emerged with:

NASA's Artemis Program (aiming for sustainable human presence).

ESA's "Moon Village" concept.

China's Chang'e missions with long-term lunar base plans.

Private companies (SpaceX, Blue Origin) envisioning lunar habitats.

## II. Moon Bases – The Architecture of Survival

A lunar base must solve three critical problems: shelter, resources, and sustainability.

### 1. Lunar Habitats

Early modules may resemble inflatable space stations or underground shelters.

Protection needed against:

- Extreme temperature swings (−173°C to +127°C).
- Micrometeorite impacts.
- Cosmic radiation and solar storms.

Possible solutions:

- Covering bases with lunar regolith (soil).
- Using 3D printing to create bricks from Moon dust.
- Building habitats inside lava tubes (ancient volcanic tunnels).

### 2. Energy Sources

Solar panels near the Moon's poles (where sunlight is constant for long periods).

Nuclear reactors for backup and night cycles.

### 3. Water and Resources

Discovery of water ice in lunar poles (confirmed by Chandrayaan-1 and other missions) is a game-changer.

Water = drinking, agriculture, and oxygen production.

Hydrogen + oxygen = rocket fuel (enabling Moon to be a "fuel station" for Mars).

### 4. Food Production

Early missions: food imported from Earth.

Long-term: hydroponics, algae bioreactors, and lunar greenhouses for self-sufficiency.

## III. International Collaboration and Competition

### 1. NASA's Artemis Program

Goal: Land humans on the Moon again (Artemis III by mid-2020s).

Establish Artemis Base Camp with habitats, rovers, and power stations.

Partnership with ESA, JAXA, and Canadian Space Agency.

### 2. China's Lunar Plans

International Lunar Research Station (ILRS) with Russia.

Target: 2030s permanent base at lunar south pole.

### 3. Europe's "Moon Village"

ESA director Johann-Dietrich Wörner proposed a global collaborative settlement.

### 4. Private Players

SpaceX: Starship for lunar cargo and tourism.

Blue Origin: "Blue Moon" lander for habitats.

Axiom, Astrobotic, ispace: supporting technologies for transport and resource utilization.

## ✈ IV. Lunar Tourism – The Next Industry

### 1. Early Experiments

2001: Dennis Tito became the first space tourist (to ISS).

Future: wealthy individuals flying around the Moon or landing for short stays.

### 2. Proposed Tourist Activities

Lunar orbital flights (SpaceX's "dearMoon" project by Yusaku Maezawa).

Short stays in pressurized habitats.

Moonwalks and rover rides.

Viewing Earthrise from lunar surface – a life-changing experience.

### 3. Economic Aspects

Tickets may cost hundreds of millions of dollars initially.

Long-term: cheaper as technology improves.

Lunar tourism could lead to luxury resorts, scientific tours, and even extreme sports (like low-gravity jumps).

## V. Challenges and Risks

Cost: Launching cargo to the Moon costs thousands of dollars per kg.

Safety: Radiation, lack of atmosphere, micrometeorites.

Medical: Effects of low gravity (1/6th of Earth).

Legal: Who owns the Moon? The 1967 Outer Space Treaty forbids territorial claims, but resource rights remain disputed.

Ethics: Should lunar landscapes be preserved like natural parks?

### VI. The Future – Moon as a Gateway

A lunar base could serve as a stepping stone to Mars.

Moon tourism may evolve into permanent settlements for scientists, miners, and eventually families.

The Moon might become:

- A science hub (astronomy, geology).
- A mining hub (helium-3 for fusion energy).
- A cultural hub (tourism, art, and even sports).

### Conclusion

Moon bases and lunar tourism represent the next great leap in human civilization. What began as dreams of storytellers is now becoming the blueprint of engineers. The Moon is not only Earth's closest neighbor but also a gateway to the stars.

Whether for science, survival, or adventure, humanity is preparing to set down roots on the Moon. The first settlers may be astronauts and scientists, but eventually, tourists, entrepreneurs, and even children might call the Moon their temporary home.

The Moon will no longer be a distant light in the night sky – it may become our second home.

# 47 - Helium-3 Mining And The Lunar Economy

## The Promise of Helium-3

The Moon is not only a celestial wonder and a potential site for future human colonies, but also a cosmic treasure chest. Among its most fascinating resources is Helium-3 ($^3He$), a rare isotope on Earth but comparatively abundant on the lunar surface.

## Why does Helium-3 matter?

It is considered the holy grail of clean energy, potentially fueling nuclear fusion reactors.

Fusion with Helium-3 releases tremendous power without radioactive waste that plagues current nuclear technologies.

If harnessed successfully, it could reshape global energy systems, geopolitics, and economics.

This chapter explores the science, feasibility, economic prospects, and geopolitical implications of Helium-3 mining on the Moon, while critically examining challenges and ethical questions.

## I. What is Helium-3?

### 1. The Isotope

Helium-3 is a non-radioactive isotope of helium (two protons, one neutron).

Fusion Reaction:

$^3He$ + $^2H$ (Deuterium) → Helium-4 + Proton + 18.4 MeV energy.

Clean energy: Unlike traditional nuclear reactions, it does not emit dangerous neutrons.

### 2. Scarcity on Earth

Naturally rare: Only ~15 kilograms exist on Earth in usable form.

Mostly produced as a byproduct of nuclear weapons programs and tritium decay.

Current cost: Estimated $40,000 per gram.

**3. Abundance on the Moon**

Solar wind has implanted Helium-3 particles into the lunar regolith (soil).

Estimated reserves:

~1 million metric tons on the Moon.

Even 25% of this could power Earth for centuries.

## II. Sources and Distribution on the Moon

**1. Deposition by Solar Wind**

Earth's magnetic field shields us from solar wind, but the Moon has no such protection.

As a result, its soil has soaked in Helium-3 over billions of years.

**2. Concentration Regions**

Higher density near the lunar maria (basalt plains).

Particularly abundant at polar regions, where the soil is older and less disturbed.

**3. Extraction Estimates**

To produce 1 ton of Helium-3, around 150 million tons of lunar soil must be processed.

This highlights both opportunity and the enormous scale of the challenge.

## ⚙ III. Mining and Extraction Techniques

**1. Regolith Processing**

Heating regolith to ~700°C releases Helium-3 gas.

Collected and compressed into storage containers.

### 2. Robotic Mining

Autonomous rovers dig and transport regolith.

Microwave heating techniques considered for energy efficiency.

### 3. Transport to Earth

Requires cryogenic storage.

Delivered using lunar landers or reusable rockets (e.g., SpaceX Starship).

## ⚡ IV. Helium-3 Fusion and Energy Potential

### 1. Advantages of Helium-3 Fusion

No radioactive waste (proton-emitting reaction).

High energy yield – 18.4 MeV per reaction.

Safer reactors (no runaway chain reactions).

### 2. Energy Yield Comparison

| Fuel Type | Energy Yield | Waste Product |
|---|---|---|
| Uranium-235 | High | Radioactive |
| Deuterium-Tritium | High | Neutrons, waste |
| Deuterium-Helium-3 | High | No neutrons |

### 3. Global Energy Replacement

25 tons of Helium-3 → could power United States for a year.

100 tons annually → could meet the entire world's needs.

## V. Economic Dimensions

### 1. Lunar Economy Potential

Helium-3 could be worth trillions of dollars.

Spinoff industries:

- Lunar construction (habitats, mining bases).
- Lunar tourism (synergy with mining infrastructure).
- Deep-space fuel depots.

### 2. Cost Challenges

Current lunar missions cost billions.

Large-scale mining may need international partnerships.

### 3. Private Sector Interest

Companies like Blue Origin, SpaceX, and ispace envision resource extraction.

Venture capital increasingly flows into space mining startups.

## VI. Geopolitics of Lunar Resources

### 1. International Treaties

Outer Space Treaty (1967): No nation can claim sovereignty over the Moon.

Artemis Accords (2020): Framework for resource utilization.

### 2. Potential Rivalries

USA, China, India, Russia, Europe – all pursuing lunar missions.

Helium-3 could trigger a new "Energy Cold War".

### 3. Resource Sharing vs Competition

Ethical dilemma: Will lunar resources benefit all humankind or only powerful nations?

## VII. Challenges and Criticisms

### 1. Scientific Uncertainty

Practical Helium-3 fusion has not yet been achieved in reactors.

Technology may take decades.

### 2. Mining Logistics

Extreme lunar conditions (vacuum, radiation, dust hazards).

High transport cost back to Earth.

### 3. Environmental Concerns

Mining could scar lunar landscapes.

Ethical debate: Should humanity exploit the Moon like Earth?

## VIII. Future Scenarios

### 1. Optimistic

By mid-21st century, Helium-3 powers clean fusion plants on Earth.

Lunar mining becomes backbone of a space economy.

### 2. Realistic

Mining begins with pilot projects.

Energy yield remains supplementary, not dominant.

### 3. Pessimistic

Fusion proves unfeasible.

Helium-3 hype fades, but lunar infrastructure supports tourism and science instead.

## Conclusion – The Moon as the Power Plant of the Future

Helium-3 offers an extraordinary promise: safe, clean, virtually limitless energy. If technology matures, lunar mining could revolutionize not only energy but also geopolitics, economics, and humanity's place in space.

But challenges remain immense: scientific, economic, ethical, and political.

The Moon may thus become both a test ground for technology and a mirror of humanity's values. Will we treat it as a shared heritage of mankind, or the next frontier of resource exploitation? The answer will define the Lunar Economy of the 21st century.

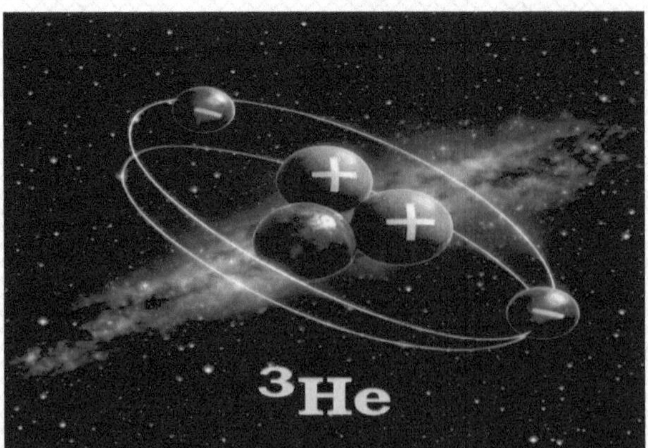

# 48 - Predictions Of The Moon's Future In Science

## Gazing Forward into Lunar Destiny

For millennia, humans have looked up at the Moon as a timeless, unchanging beacon. Yet, science shows us that the Moon is not static. It is a dynamic celestial body, gradually changing in orbit, rotation, and structure. The Moon has a past written in craters and volcanic plains, but it also has a future shaped by gravitational interactions, internal cooling, and human exploration.

This chapter explores scientific predictions about the Moon's fate over millions and billions of years—from orbital drift and geological changes to human colonization and potential collapse scenarios. By combining astronomy, geology, space engineering, and futuristic visions, we can begin to imagine what the Moon of tomorrow might look like.

## I. Orbital Evolution of the Moon

### 1. Receding Moon

The Moon is moving away from Earth at ~3.8 cm per year.

This was first confirmed by Apollo laser reflectors placed on the surface.

In ~600 million years, a total solar eclipse will no longer be possible, as the Moon will appear too small to cover the Sun.

### 2. Tidal Lock Stability

Currently, the Moon is tidally locked—showing the same face to Earth.

Predictions show this lock will remain stable for billions of years.

However, Earth's slowing rotation and the Moon's drift will alter the length of a day → from 24 hours today to ~47 hours in 5 billion years.

### 3. Long-Term Earth-Moon System Fate

As the Sun evolves into a red giant (~5 billion years from now), the Earth-Moon system may destabilize.

The Moon might spiral outward further or be engulfed along with Earth.

## II. Geological Future of the Moon

### 1. Cooling Interior

The Moon's core is still partially molten, but it is slowly cooling.

Over the next few billion years → seismic activity ("moonquakes") will decline further.

### 2. Space Weathering

Constant micrometeorite bombardment and solar wind will continue to darken and erode the regolith.

The surface will grow smoother and dustier.

### 3. Crater Accumulation

With no atmosphere, new craters will keep forming indefinitely.

In the next billion years, the Moon's surface will be even more scarred, preserving the "cosmic history book."

## III. Lunar Water and Ice Predictions

### 1. Polar Ice Stability

Ice trapped in permanently shadowed craters near the poles could survive for billions of years.

Human colonization may consume some of these reserves.

### 2. Future Migration of Volatiles

Comet impacts or solar activity may redistribute water over time.

Some predictions suggest creation of localized thin exospheres of vapor after impacts.

## IV. Human Influence on the Moon's Future

### 1. Colonization Scenarios

By 21st–22nd century, permanent lunar bases may become reality.

Colonies could expand near polar regions with access to ice and sunlight.

Long-term → entire lunar cities, agriculture domes, and industries may reshape the Moon's landscape.

### 2. Industrial Mining

Extraction of helium-3, rare metals, and water will leave visible scars.

Predictions suggest that by 2200 CE, the Moon could host a lunar economy supporting Earth.

### 3. Terraforming Ideas

Some scientists propose installing artificial magnetic shields or solar reflectors to stabilize the lunar environment.

Although speculative, such interventions could redefine the Moon's natural history.

## V. Scientific Predictions of Long-Term Fate

### 1. In the Next 1,000 Years

Moon bases, mining, and tourism reshape the surface.

Laser communications and giant telescopes dominate the far side.

### 2. In the Next 1 Million Years

The Moon drifts ~38 km farther from Earth.

Total eclipses end; partial eclipses continue.

Surface pockmarked with new craters.

### 3. In the Next 1 Billion Years

Earth days lengthen beyond 30 hours.

Moon ~40,000 km farther away.

Human impact may vanish if civilizations collapse, but natural geology continues.

### 4. In the Next 5 Billion Years

Sun enters red giant phase.

Earth may be engulfed, possibly pulling the Moon along.

Alternative scenario: Moon escapes and orbits the dying Sun independently.

## VI. Cultural and Psychological Predictions

The Moon will always remain a symbol in human imagination.

Even if colonized, it will inspire poetry, mythology, and identity.

Predictions suggest new "lunar mythologies" will arise around human settlers—similar to how ancient Earth cultures wove myths around the Moon.

## Conclusion – The Eternal Yet Changing Moon

The Moon is not a static object but a living partner in Earth's cosmic dance. Science predicts:

It will drift farther, altering eclipses and Earth's tides.

Its geology will evolve into a smoother, dustier, more cratered landscape.

Human colonization may leave lasting technological signatures on its face.

Ultimately, whether engulfed by the Sun or wandering freely in space, the Moon's story will always be intertwined with humanity. From myth to science, from stone age rituals to futuristic mining, the Moon remains both our mirror of the past and a beacon of the future.

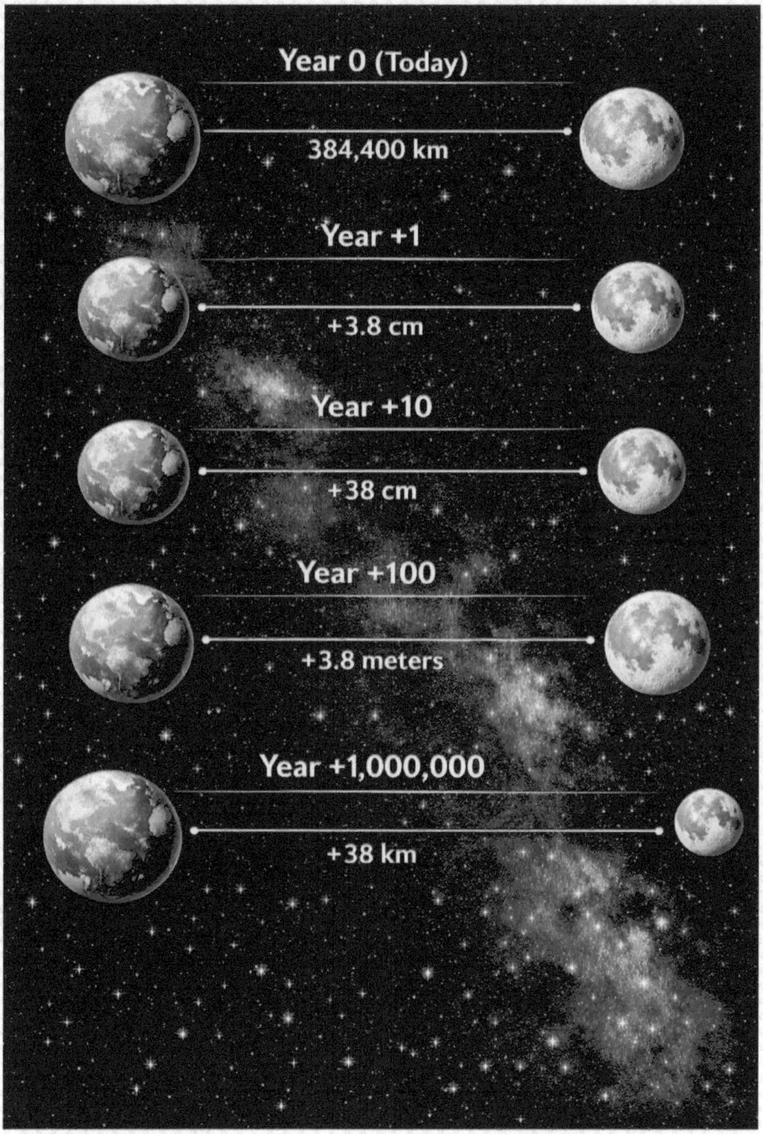


**Moon drifting away over time.**

# 49 - Moon In Science Fiction And Futurism

## The Moon as Imagination's Playground

From the earliest tales of gods and spirits to the modern age of rockets and robotics, the Moon has remained humanity's cosmic mirror—not only reflecting light, but also reflecting our deepest fears, hopes, and dreams. In science fiction and futurism, the Moon has consistently served as a stage upon which humanity projects its visions of conquest, colonization, apocalypse, and utopia.

This chapter explores the Moon's role in science fiction literature, films, and futurist thought—how writers, scientists; and dreamers imagined moon travel long before Apollo, how the Moon continues to be central in speculative fiction, and how futuristic scenarios like lunar cities, economies, and post-human societies are evolving.

## I. Early Literary Imaginations

1. Classical and Medieval Fantasies

Lucian of Samosata (2nd century CE) – *True History* described a fantastical voyage to the Moon, where alien civilizations thrived.

Medieval Islamic scholars such as Al-Farghani speculated about celestial travel, indirectly inspiring moon-journey allegories.

2. Renaissance to Enlightenment

Johannes Kepler's Somnium (1609, published 1634) – often called the first true science fiction, describing how Earth might appear from the Moon.

Cyrano de Bergerac's "Voyages to the Moon" (1657) – whimsical, but introduced mechanical devices and rockets to reach the Moon.

3. 19th Century Proto-Sci-Fi

Jules Verne's "From the Earth to the Moon" (1865) – a cannon-propelled space capsule sent explorers to the Moon. His remarkable technical details foreshadowed Apollo missions.

H.G. Wells' "The First Men in the Moon" (1901) – introduced the insect-like "Selenites," exploring not just travel, but alien societies beneath the lunar crust.

## II. The Moon on Screen

1. Early Films

Georges Méliès' "A Trip to the Moon" (1902) – iconic silent film where explorers fire themselves from a cannon, land in the Moon's eye, and battle lunar creatures.

This film became a visual archetype of the Moon as both comic and mysterious.

2. Cold War & Space Race Cinema

Hollywood films like *Destination Moon (1950)* emphasized scientific accuracy.

Soviet cinema (*Road to the Stars*, 1957) portrayed the Moon as a stepping-stone to interplanetary colonization.

3. Post-Apollo Narratives

Once man reached the Moon, films began to imagine lunar colonies, disasters, and hidden alien bases.

*2001: A Space Odyssey* (1968) showed a lunar monolith, tying the Moon to cosmic intelligence.

*Moon* (2009) by Duncan Jones explored psychological isolation of lunar miners.

## III. Themes in Moon Science Fiction

| Theme | Examples | Significance |
|---|---|---|

| Exploration | Verne, Wells, Méliès | Early dreams of reaching Moon |
|---|---|---|
| Alien Life | Wells' *Selenites*, pulp fiction aliens | Moon as a habitat for Others |
| Colonization | Clarke's *Earthlight*, Heinlein's *The Moon is a Harsh Mistress* | Moon as a second Earth |
| Resource Extraction | Heinlein, modern Helium-3 stories | Foreshadowing lunar economy |
| Psychological Isolation | *Moon* (2009), *Apollo 18* (2011) | Human fragility in alien worlds |
| Myth & Magic Reimagined | Fantasy novels, anime, manga | Moon as mystical or divine realm |

## IV. The Moon in Futurism

1. Post-Apollo Futurism

Futurists in the 1970s envisioned:

- Lunar cities with geodesic domes.
- Space elevators connecting Earth and Moon.
- Industrial hubs exporting resources back to Earth.

2. Helium-3 and Energy Utopias

Writers and scientists alike saw the Moon as a potential energy solution for Earth through nuclear fusion.

Science fiction often imagines wars or corporate empires fighting over lunar Helium-3.

3. Political Futures

Scenarios of a lunar Cold War between superpowers.

Heinlein's *The Moon is a Harsh Mistress* introduced the idea of a lunar independence revolution.

## V. Moon in Cyberpunk and Post-Human Futures

In cyberpunk and post-human fiction, the Moon often becomes:

A prison colony (used in dystopian narratives).

A luxury resort for the ultra-rich.

A gateway hub for asteroid mining and Mars missions.

A digital-immortality storage site (AI minds housed in underground lunar bases).

Example: In anime like *Cowboy Bebop* and *Mobile Suit Gundam*, the Moon is central to interplanetary politics and battles.

## VI. Futuristic Scenarios

1. Near Future (21st–22nd Century)

Permanent research stations.

Tourism for billionaires.

Lunar resource exports (water, metals, Helium-3).

2. Mid Future (23rd–25th Century)

Fully functioning lunar cities.

Bio-engineered humans adapted to low gravity.

Cultural divergence → "Lunarians" with distinct identity.

3. Far Future (Beyond 3000 CE)

Moon turned into a giant observatory.

Potential transformation into a Dyson-like station orbiting Earth.

Or even destroyed in a cosmic accident or human exploitation.

### Conclusion – The Moon as a Future Mirror

The Moon in science fiction and futurism is not just about space travel—it is about human destiny.

Will we exploit it as a resource hub?

Transform it into a new civilization?

Or destroy it in our greed?

In every future scenario, the Moon remains the canvas of imagination.
It is both real rock and poetic dream, a place where science, myth, and human ambition intersect.

Méliès' Moon with a rocket in its eye – symbol of early sci-fi wonder.

# 50 - Return To The Moon – Artemis, Isro & Beyond

## The Moon Calls Again

More than fifty years after Neil Armstrong's first step, humanity is preparing for a new era of lunar exploration. The Moon is no longer seen only as a symbol of achievement, but as a strategic gateway to deeper space, a potential hub for resources, and a stage for international cooperation—and competition.

The current wave of missions, led by NASA's Artemis program, India's Chandrayaan and upcoming ISRO missions, China's Chang'e program, Japan's SLIM, and private ventures like SpaceX and Blue Origin, represent a multi-polar return to the Moon. Unlike Apollo, this era is about sustainability, science, and future settlement.

This chapter explores Artemis, ISRO's vision, global collaborations, private sector involvement, and what "beyond" may look like in the decades ahead.

## I. NASA's Artemis Program – Back to Stay

### 1. Origins and Goals

Artemis named after Apollo's twin sister in Greek mythology, symbolizing a return to the Moon with diversity and sustainability.

Core objectives:

Land the first woman and first person of color on the Moon.

Establish sustainable lunar presence by late 2020s.

Prepare technologies for Mars exploration.

### 2. Artemis Missions

Artemis I (2022) – Uncrewed flight of the Orion spacecraft around the Moon. Successfully demonstrated the Space Launch System (SLS).

Artemis II (planned 2026) – First crewed mission; astronauts will orbit the Moon without landing.

Artemis III (planned ~2026–27) – First crewed landing at the lunar South Pole.

Artemis IV–V – Construction of Gateway lunar station in orbit and multiple landings with reusable systems.

### 3. The Lunar Gateway

A modular space station orbiting the Moon, serving as:

Science laboratory.

Docking hub for landers and Orion.

Testing ground for Mars technologies.

### 4. International Contributions

ESA – Orion service module and Gateway modules.

JAXA – Life-support systems, lunar rover plans.

CSA (Canada) – Robotic arm (Canadarm3).

## II. ISRO and India's Lunar Ambitions

### 1. Chandrayaan Legacy

Chandrayaan-1 (2008) – Orbiter that confirmed water molecules on the Moon.

Chandrayaan-2 (2019) – Orbiter successful; Vikram lander crash-landed.

Chandrayaan-3 (2023) – Historic: India became the first country to land near the Moon's South Pole. Rover "Pragyan" studied soil, confirmed presence of sulfur.

### 2. Future ISRO Missions

LUPEX (Lunar Polar Exploration Mission) with JAXA – Planned late 2020s; rover designed to study water ice deposits.

Human Spaceflight Vision (Gaganyaan → Lunar) – By 2035, ISRO plans a crewed lunar mission, potentially aligning with Artemis collaborations.

### 3. India's Role in Global Collaboration

India is a signatory of the Artemis Accords.

ISRO focusing on cost-effective lunar technology, making India a leader in affordable exploration.

## III. China's Chang'e Program – The Rival Superpower

1. Achievements

Chang'e 3 (2013) – First soft landing after decades.

Chang'e 4 (2019) – First landing on the far side of the Moon.

Chang'e 5 (2020) – Sample return mission; brought lunar rocks back to Earth.

Chang'e 6 (2024–25) – More sample returned from South Pole-Aitken Basin.

### 2. Future Plans

Chang'e 7 & 8 (late 2020s) – Building blocks for a permanent lunar research base with Russia.

China envisions a "International Lunar Research Station" (ILRS) by the 2030s—potentially rivaling Artemis.

## IV. Other Global & Private Efforts

### 1. Japan – JAXA

SLIM (2023) – Precise landing technology ("Moon Sniper"), succeeded in pinpoint landing.

LUPEX (with ISRO) – Exploring polar ice.

**2. Russia**

Luna-25 (2023) – Attempted landing but crashed.

Plans to join China's ILRS initiative.

**3. Europe**

ESA contributing to Artemis Gateway, and exploring robotic mining technology.

**4. Private Ventures**

SpaceX Starship HLS – Selected as Artemis lander.

Blue Origin's Blue Moon lander – Competing in NASA contracts.

Private tourism missions (e.g., SpaceX's dearMoon project).

## V. Why the Moon Matters Again

| Aspect | Importance |
|---|---|
| Science | South Pole water ice, geology, lunar history. |
| Economy | Helium-3, rare earths, in-situ resource utilization (ISRU). |
| Technology | Testing habitats, rovers, life-support for Mars. |
| Geopolitics | Strategic control, global leadership in space. |
| Culture | Renewed inspiration, tourism, and human expansion. |

## VI. Beyond Artemis and ISRO – Future Visions

**1. Permanent Lunar Settlements**

By 2040, possible human habitats at South Pole (powered by sunlight on peaks of eternal light).

**2. Lunar Industry**

Mining of regolith, water extraction, 3D-printing of habitats.

Factories on the Moon producing fuel for Mars missions.

**3. Tourism and Colonization**

Private companies offering stays in lunar hotels.

Long-term prospect: Chandra as the "8th continent" of Earth.

**4. Toward Mars**

Lunar bases as launchpads for interplanetary travel.

"Moon to Mars" pipeline: Artemis → Gateway → Mars mission by 2040s.

## Conclusion – Humanity's Second Giant Leap

The 21st century Moon missions are not just about repeating Apollo's glory. They are about staying, building, and expanding.

Artemis symbolizes diversity and international partnership.

ISRO embodies affordability and innovation.

China's Chang'e asserts strategic dominance.

Private companies open new economic frontiers.

Together, these forces are shaping a multi-polar lunar future—a return not for a single flag, but for all of humanity.

The Moon is no longer just a destination. It is becoming the launchpad of humanity's cosmic destiny.

Moon to Mars!
Artemis Base
Gateway
Mars Mission
Artemis → Gateway → Mars Mission by 2040s